AF263737

THE SILENT EPIDEMIC Death
by Hospital Negligence

Dr. Dorathy U. Ubiam

Author Bio

Dr. Dorathy U. Ubiam is a #1 *New York Times* bestselling author and an accomplished healthcare administration expert with a Doctor of Healthcare Administration (DHA) from Walden University. With over years of experience in long-term care and healthcare systems management, she has built a reputation for transforming complex administrative challenges into strategic, results-driven solutions that improve patient outcomes and organizational performance.

Dr. Ubiam specializes in healthcare quality improvement, regulatory compliance, operational efficiency, and patient safety advocacy. Her work bridges the gap between frontline care delivery and executive-level decision-making, equipping healthcare leaders with practical frameworks to enhance accountability, reduce risk, and strengthen institutional integrity.

As an author, Dr. Ubiam is known for her insightful analysis and compelling ability to address critical issues within modern healthcare systems. Her writing reflects both scholarly depth and real-world experience, offering readers clarity, awareness, and actionable guidance. Through her work, she advocates for transparency, ethical leadership, and systemic reform in healthcare institutions.

Beyond her professional achievements, Dr. Ubiam is deeply committed to family and community. She lives with her husband and two children, who inspire her daily dedication to advancing safer, more efficient healthcare systems for future generations.

Table of Contents

CHAPTER ONE

The Empty Bed

Dr. Dorathy U. Ubiam

"The measure of a society is found in how it treats its weakest and most helpless citizens."

— Jimmy Carter

The bed in room 417 was freshly made. The sheets were pulled taut at the corners, the pillow fluffed and centered with the kind of geometric precision that only comes when someone will never lie there again. A laminated card on the bedside table still read *David M. — Knee Arthroscopy — Dr. Patel*, but David was gone. He had been gone for thirty-six hours, though the paperwork had taken longer than his dying. A nurse would strip the card by afternoon, and by evening, another patient would occupy the space, blissfully unaware of what had happened in the silence between midnight and dawn.

Outside the room, the hallway hummed with its usual choreography. A phlebotomist pushed a cart toward the elevator. Two residents compared notes over a vending machine lunch. A family in the waiting area scrolled through their phones, waiting for news about someone else's surgery. The machinery of the hospital continued to turn, as it always does, because hospitals are designed to absorb death the way rivers absorb rain—quietly, without interruption, without altering the current. There would be no announcement over the intercom. No moment of collective silence. David's death was, in the language of hospital administration, a "sentinel event." It would be logged, reviewed, and filed. The word "sentinel" comes from the Latin *sentire*—to feel, to perceive. The irony is almost unbearable. The system named the event after the very faculty it had failed to exercise.

This is the story the hospital would rather not tell. Not because it is extraordinary, but because it is ordinary. It happens in fluorescent-lit corridors across the country, in community hospitals and research institutions alike, in places where the coffee is always burnt, and the overhead pages never stop. It happens so often, in fact, that we have given it a sterile, bureaucratic name: adverse event. As if the word "event" could contain the weight of a family's collapse.

David's wife, Maria, received the call at 3:47 in the morning. She would later tell a journalist that the first thing she noticed was the caller's tone—not panicked, not emotional, but measured and procedural, as though reading from a template. "There has been a complication," the voice said. Maria drove to the hospital in her pajamas, running two red lights on empty streets. When she arrived, David was on a ventilator. His skin had already taken on the grayish pallor that nurses recognize, but families do not—the complexion of a body whose systems are shutting down one by one, like lights being switched off in a building at closing time. He never regained consciousness.

David was forty-two. He coached his daughter's soccer team on Saturdays. He was the kind of man who memorized his coworkers' birthdays and brought homemade banana bread to the office on Mondays. He went into the hospital on a Tuesday for routine knee surgery—the kind of procedure so common that surgeons sometimes compare it to an oil change. He never came home.

His daughter, who was eleven at the time, had made a get-well-soon card with a soccer ball drawing on the front. It was sitting on the kitchen counter when Maria got the phone call. The pre-operative instructions had told David to bring loose-fitting clothing for the ride home. Those clothes were still folded in a bag by the front door three weeks later, untouched, because no one in the family could bring themselves to move them. This is the texture of loss that hospital incident reports never capture—the card that was never delivered, the sweatpants that were never worn, the ordinary objects that become relics of a future stolen in the dark while monitors beeped in an empty hallway.

What killed David was not a scalpel that slipped or a drug that was mislabeled. What killed David was silence. Specifically, the silence of an alarm that rang so often, and so meaninglessly, that the human ears responsible for hearing it had long since learned to tune it out.

The Quiet Architecture of Negligence

When most people hear the phrase "hospital negligence," they picture something dramatic: a surgeon amputating the wrong limb, a nurse injecting a lethal dose, a catastrophic and unmistakable blunder. Hollywood has trained us to think of medical error as a single, decisive moment—a villain in scrubs, a smoking gun. But the truth is far more unsettling. The truth is mundane.

Hospital negligence, in its most lethal form, is a series of small, ignored flickers. It is a lab result that arrives during a shift change and falls between the cracks of two exhausted physicians' attention. It is a monitor alarm that blares so frequently—dozens of times per hour in a busy ICU—that the staff has developed a clinical term for their indifference: alarm fatigue. It is a nurse working her third consecutive twelve-hour shift, too depleted to notice that a patient's breathing has subtly changed from rhythmic to labored. It is not one failure. It is a cascade.

Consider the sheer volume of decisions a hospital generates in a single day. A mid-sized community hospital with 300 beds will process, on average, thousands of medication orders, hundreds of diagnostic tests, dozens of shift handoffs, and countless verbal instructions exchanged among physicians, nurses, pharmacists, and technicians. Each of these transactions is an opportunity for precision. Each is also an opportunity for error. The difference between the two often comes down to factors unrelated to the competence of the individuals involved: Was the electronic health record system functioning properly during the handoff? Was the nurse-to-patient ratio low enough to allow attentive monitoring? Was the alarm threshold calibrated to the patient's specific baseline, or was it set to a factory default designed for a hypothetical average that matches no real person?

The scale of the problem defies intuition. According to a landmark study published in the British Medical Journal, medical errors are the third leading cause of death in the United States, claiming more than 250,000 lives each year. A subsequent analysis by researchers at Johns Hopkins placed the figure even higher, estimating more than 440,000 annual deaths attributable to preventable harm in hospitals. To put the lower number in perspective, it is equivalent to two fully loaded commercial airplanes crashing every day, with no survivors, for an entire year. If that were happening in aviation, the industry would be shut down within a week. In healthcare, it is simply the cost of doing business.

Yet these numbers, staggering as they are, almost certainly undercount the true toll. Death certificates rarely list "medical error" as a cause of death. The International Classification of

Diseases, which governs how deaths are coded in the United States, has no specific category for it. A patient who dies of cardiac arrest triggered by an undetected medication interaction is coded as a cardiac death, not a negligence death. A grandmother who suffers a fatal heart attack at home after being wrongly discharged from the ER is coded as a myocardial infarction, not a diagnostic failure. The data itself is designed to obscure the very patterns it should reveal.

The numbers are staggering, but numbers alone cannot convey the devastation. Behind each data point is a room like 417—a freshly made bed and a family trying to comprehend how a routine procedure, a minor inconvenience on a calendar, became the last entry in someone's life.

Why This Book Exists

This book was not written to demonize doctors or nurses. The vast majority of healthcare workers are deeply dedicated professionals who entered the field to heal. Many of them carry the weight of patient losses with a grief that rivals that of the families they serve. They work punishing hours in systems that treat their labor as expendable and their burnout as a personal failing. This book is not about them as individuals. It is about the machinery they are trapped inside.

There is a term in patient safety research that most people outside the field have never heard: the "second victim." It refers to the healthcare provider involved in a medical error— the nurse who realizes, hours later, that the alarm she dismissed was real; the physician who discovers that the lab result he never saw could have saved a life. Studies suggest that as many as half of all healthcare workers will experience a second-victim event during their careers. Many suffer from depression, anxiety, and post-traumatic stress. Some leave the profession entirely. A disturbing number contemplate suicide. And yet, the institutional response to these workers is often silence or discipline—the same institutional silence that failed their patients in the first place. The system creates victims on both sides of the bed and then refuses to acknowledge either.

The central argument of this book is straightforward, though its implications are profound: death by negligence is a systemic failure, not merely an individual one. When a patient dies because an alarm was ignored, the instinct of hospital administrators, legal teams, and regulatory bodies is to find the person who failed—the nurse who didn't respond, the physician who didn't check. This approach is not only insufficient; it is dangerous. It treats the symptom while the disease metastasizes.

The nurse who ignored the alarm was on her fourteenth consecutive hour. The staffing model that put her there was designed by administrators who prioritized cost efficiency over patient safety. The alarm system itself was configured with factory-default thresholds that triggered so many false positives that real emergencies were drowned out in a sea of electronic noise. The physician who missed the critical lab result was using software with a known glitch during shift transitions—a glitch that had been reported to IT three times and had been deprioritized each time.

Each of these failures, taken alone, is survivable. Taken together, they are lethal. This is the architecture of negligence: not a single crack, but a network of fractures that, under pressure, gives way all at once.

Toward Radical Accountability

The purpose of this book is not merely to catalog failures. Cataloging failures is easy; it is the default mode of investigative journalism and malpractice litigation. What is far harder—and far more necessary—is to propose a new framework for understanding and preventing these failures. That framework is what I call Radical Accountability.

Radical Accountability begins with a simple but disruptive question: instead of asking *"Who is to blame?"* we ask *"How did the system fail them?"* This is not a softer question. It is a harder one. It demands that we trace the roots of every preventable death not to a single person's lapse in judgment, but to the institutional, technological, and cultural conditions that made that lapse inevitable.

Radical Accountability means that when a patient dies because an alarm was ignored, we do not stop at the nurse's disciplinary hearing. We redesigned the alarm system. We overhaul the staffing model. We create a culture where reporting near-misses is rewarded rather than punished. We borrow from industries—aviation, nuclear power, automotive manufacturing—that have already proven that systemic thinking can drive error rates to near zero.

It means acknowledging an uncomfortable truth: that the healthcare system, as currently designed, is not optimized for patient safety. It is optimized for throughput, for revenue, for liability management. Patient safety is a byproduct, not a design principle. Radical Accountability demands that we reverse this hierarchy.

The aviation industry offers a powerful proof of concept. In the 1950s, commercial air travel was plagued by catastrophic accidents, many of them caused by human error compounded by systemic flaws. The industry's response was not to fire more pilots. It was to redesign the entire flight ecosystem. Cockpit resource management protocols were introduced to ensure that co-pilots could challenge captains without fear of retribution. Black box recorders were mandated so that every failure could be studied forensically. A non-punitive reporting culture was established through programs that encouraged pilots and crew to report near-misses and safety concerns without threat of disciplinary action. The result was a transformation so complete that commercial aviation is now the safest form of mass transportation on earth. The healthcare industry, by contrast, remains stuck in the 1950s model: find the person, assign the blame, move on.

There is also a financial dimension to this crisis that cannot be ignored. Preventable medical errors cost the American healthcare system an estimated \$17–\$29 billion annually in direct costs alone—additional treatments, extended hospital stays, malpractice settlements, and legal fees. When indirect costs are factored in—lost productivity, disability, and the long-term healthcare needs of patients who survive errors but are permanently harmed—the figure climbs into the hundreds of billions. This is not merely a moral failure. It is an economic one. The system is bleeding money at every point where it is bleeding patients, and yet the institutional response continues to be reactive rather than preventive, punitive rather than structural.

Humanizing the Data

Throughout this book, you will encounter statistics. You will learn that diagnostic errors account for roughly 17% of adverse events in hospitals, that communication breakdowns during shift changes contribute to an estimated 80% of serious medical errors, and that the average ICU nurse is exposed to more than 350 alarms per day. These numbers matter. They provide the evidentiary scaffolding for systemic reform.

But this book refuses to let the data stand on its own. Behind every percentage is a person. Behind every "adverse event" is a David—a father who coached soccer, who made banana bread, who walked into a hospital expecting to walk out. Behind every "communication breakdown" is an Elena—a grandmother whose chest pain was dismissed as acid reflux because a software glitch swallowed her lab results during a shift change.

You will meet these people in the pages that follow. You will learn their names, their routines, the small rituals that defined their days. And through their stories, you will come to understand what the data alone cannot convey: that every preventable death in a hospital is not a statistic. It is an amputation of the future—every birthday unattended, every conversation unfinished, every grandchild who will know them only from photographs.

This approach is deliberate. There is a well-documented phenomenon in behavioral psychology known as the "collapse of compassion"—the finding that human beings actually become less empathetic as the number of victims increases. One death is a tragedy; a million deaths is a statistic. The phrase, commonly attributed to Joseph Stalin, carries more truth than cynicism. Psychologist Paul Slovic has demonstrated that our capacity for empathy peaks when there is a single identified victim and diminishes rapidly as numbers grow. Show someone a photograph of one starving child, and they will donate generously. Show them a photograph of two starving children, and their donations actually decrease. The mind cannot hold a quarter of a million deaths in focus. It can hold one face, one name, one story.

That is why this book insists on names. Not pseudonyms chosen for anonymity, but names chosen to restore identity to people whom the system reduced to case numbers. Every patient in these pages was someone's child, someone's partner, someone's reason for coming home. The medical record may include a diagnosis code and a date of discharge. This book remembers them as human beings. It is the least we owe them, and perhaps the most powerful thing we can offer: the simple, radical act of refusing to let them disappear.

The Questions That Demand Answers

At its core, this book is driven by three questions—the healthcare industry has, for too long, either avoided or answered with deflection. They are not rhetorical. They are urgent. And they deserve honest, unflinching answers.

First: How many preventable deaths occur in hospitals each year—and why don't we know the exact number? The fact that we cannot produce a precise, universally agreed-upon figure is itself an indictment. Estimates range from 250,000 to over 440,000 annually in the United States, depending on methodology and the willingness of institutions to report. The gap between these numbers is not a rounding error. It is a cover story. It represents the distance between what hospitals acknowledge and what actually happens behind closed doors. A system that cannot accurately count its own casualties has not yet decided to take them seriously.

Second: Why does the healthcare system persist in treating negligence as an individual failing rather than a systemic one? There is a reason the first instinct, when a patient dies from a preventable error, is to find a person to blame. Individual blame is tidy. It satisfies the legal department. It reassures the public that the "bad apple" has been removed and the barrel is clean. But the barrel is not clean. The barrel is the problem. When we punish individuals without reforming the systems that set them up to fail, we guarantee that the same errors will recur with different names and different faces. The nurse who is fired for missing an alarm is replaced by another nurse who will work the same impossible hours, in the same understaffed unit, listening to the same avalanche of meaningless beeps. Nothing changes except the name on the incident report.

Third: What would a culture of "Radical Accountability" look like in practice? This is the question that animates the second half of this book. Radical Accountability is not an abstraction. It is a set of concrete, implementable strategies drawn from industries that have already solved the problem of preventable human death at scale. It means redesigning alarm systems so that critical alerts cannot be drowned out by noise. It means staffing models that treat nurse fatigue as a patient safety variable, not a budget line to be minimized. It means closed-loop communication protocols that make it structurally impossible for a critical lab result to vanish into a digital void. And it means a cultural transformation in which reporting errors and near-misses is not an act of whistle-blowing, but a routine and celebrated part of professional practice.

These three questions form the spine of everything that follows. Every case study, every data point, every mitigation strategy in this book is in service of answering them—not with platitudes, but with evidence, and not with blame, but with blueprints for change.

What Lies Ahead

The chapters that follow are organized to move from understanding to action. We begin by dissecting the anatomy of medical error—the diagnostic failures, communication breakdowns, and technological over-reliance that underpin most hospital negligence. We examine not just what goes wrong, but why it goes wrong, tracing each failure to its systemic roots. You will learn how cognitive biases lead experienced physicians to anchor on an initial diagnosis and dismiss contradictory evidence. You will see how the "Silo Effect"—the tendency of hospital departments to operate as isolated units—creates dangerous gaps in information flow. And you will understand why the same technology that was designed to save lives can, when poorly implemented, become the very instrument of their loss.

From there, we turn to case studies—real stories of real patients whose deaths illuminate, with painful clarity, the gap between what our healthcare system promises and what it delivers. These are not cautionary tales designed to frighten. They are forensic narratives designed to educate. Each case study follows a consistent structure: the patient, the incident, the outcome, and the mitigation lesson. This structure is deliberate. It allows us to honor the person at the center of each story while also extracting the systemic insight that can prevent the next tragedy.

In the latter half of the book, we shift from diagnosis to treatment. We explore the mitigation frameworks that have the potential to transform patient safety: the Swiss Cheese Model, borrowed from accident theory; the Stop the Line authority, adapted from manufacturing; and closed-loop communication protocols that have already proven effective in aviation. We examine how artificial intelligence and real-time surveillance can catch the warning signs that exhausted human eyes miss. And we lay out a concrete Administrative Policy Roadmap—a practical blueprint that hospital leaders, patient safety officers, and policymakers can begin implementing today.

And finally, we turn to you—the patient, the family member, the advocate. Because the most underutilized safety mechanism in any hospital is the person sitting at the bedside. We will equip you with the language, the strategies, and the confidence to speak up when something feels wrong, without fear of being labeled "difficult."

A note on language. This book is written for a general audience. While it addresses complex medical, technological, and policy concepts, it translates them into accessible terms. Wherever a technical term is introduced, it is defined and illustrated with real-world examples. A glossary of key terms is provided at the end of the book for reference. This is not a textbook. It is a call to arms—and calls to arms must be understood by everyone who answers them.

The Bed in Room 417

I return to the empty bed because it is the image that should haunt us. Not the surgical theater, not the emergency room, not the dramatic moments that make for compelling television. The empty bed. The one that was occupied yesterday by someone who trusted the system to keep them safe, and whose trust was betrayed not by malice, but by indifference—by a system that had grown so accustomed to its own failures that it could no longer see them.

In the weeks after David's death, his wife Maria would drive past the hospital on her way to work. She told a reporter that she sometimes found herself staring at the windows, trying to locate the fourth floor and identify room 417 from the outside. She never could. From the street, all the windows looked the same—rows of identical glass, sealed and reflecting the sky, revealing nothing of what happened behind them. That, she said, was the thing that haunted her most. Not the phone call at 3:47 in the morning, not the ventilator, not the funeral. It was the hospital's impenetrable normalcy. The way it simply went on. The way it absorbed her husband's death and continued functioning as if nothing had happened, because, from the institution's perspective, nothing unusual had.

This book is an act of seeing. It is an attempt to make visible what the healthcare industry has rendered invisible: the quiet, incremental, devastatingly preventable ways in which hospitals fail the people they are meant to protect. It is written for the families who have been told that their loved one's death was a "complication" rather than a consequence. It is written for the nurses and physicians who carry the guilt of systemic failures they did not create but are forced to embody. And it is written for every patient who has ever lain in a hospital bed, trusting that the beeping of the monitors meant someone was watching, that the silence in the hallway meant all was well.

David's bed has been remade. The laminated card has been replaced. But somewhere, in a hospital not far from where you are reading this, another alarm is sounding. The question this book asks—the question *all of us* must ask—is simple:

Is anyone listening?

CHAPTER TWO

The Anatomy of a Medical Error

Dr. Dorathy U. Ubiam

"To err is human; to cover up is unforgivable; to fail to learn is inexcusable."

— Sir Liam Donaldson, Former Chief Medical Officer of England

A medical error does not announce itself. There is no siren, no flashing light, no moment where the room falls silent and everyone looks at each other in dawning horror. In the movies, medical catastrophe is a spectacle—a flatline on a monitor, a surgeon's trembling hands, a shouted code blue that sends a team sprinting down the corridor. In reality, medical errors are almost always invisible at the point of occurrence. They are quiet. They are procedural. They slip into the machinery of care like a grain of sand into a gearbox—unnoticed at first, and then, by the time the grinding begins, already too late to extract.

Understanding how medical errors happen requires us to abandon the comforting fiction that they are caused by bad people making bad decisions. The overwhelming body of evidence in patient safety research points to a different conclusion: most medical errors are committed by competent, well-intentioned professionals operating within systems that are designed to fail. The error is the final, visible symptom of a disease that lives deep in the institutional structure—in the way information moves (or fails to move) between departments, in the cognitive shortcuts that exhausted minds are forced to take, in the technological tools that promise precision but deliver fragility.

This chapter dissects three of the most common and most lethal categories of medical error: diagnostic failure, communication breakdown, and technological over-reliance. Each operates through its own distinct mechanism. Each has its own body count. And each, as we will see, is eminently preventable—if the system can summon the will to change.

Diagnostic Failure: Why Doctors Miss the Obvious

Imagine you are an emergency room physician. It is eleven o'clock on a Friday night. You have been on your feet for nine hours. You have seen forty-three patients. A man walks in complaining of chest tightness and fatigue. He is thirty-eight, slightly overweight, and admits to having eaten a large meal two hours ago. He has no history of cardiac disease. His EKG is normal. His vital signs are stable. Every piece of available data points to the same conclusion: gastroesophageal reflux—heartburn.

You prescribe an antacid and send him home. Twelve hours later, he is dead of a massive heart attack. The autopsy reveals severe coronary artery disease that had been developing for years, silently, asymptomatically, until it wasn't. The question that will haunt the malpractice review, the hospital inquiry, and your own sleepless nights for years to come is: How did you miss it?

The answer, in most cases, is not incompetence. It is cognition. Specifically, it is a set of cognitive biases so deeply wired into the human brain that even the most experienced and well-trained physicians fall prey to them, often without realizing it. Understanding these biases is essential to understanding why diagnostic errors remain one of the most persistent and deadly categories of medical failure.

The first and most pervasive is **anchoring bias**—the tendency to fixate on the first piece of information encountered and to interpret all subsequent information through that initial lens. In the scenario above, the physician anchored on the patient's recent large meal and his lack of cardiac history. These two facts created a frame—heartburn—and once that frame was established, every subsequent data point was unconsciously filtered to fit it. The normal EKG confirmed the frame. The stable vitals confirmed the frame. The patient's relatively young age confirmed the frame. What the physician did not do—what the anchor prevented him from doing—was ask the one question that might have shattered the frame: What if this isn't heartburn? What else could it be?

Closely related to anchoring is **availability bias**—the tendency to judge the likelihood of a diagnosis based on how easily similar cases come to mind. A physician who has seen twenty cases of acid reflux in the past week and zero cases of myocardial infarction in patients under forty will unconsciously assign a lower probability to heart attack, not because the medical evidence supports that assessment, but because the brain's statistical software is weighted by recency, not by reality. The most recent experience becomes the most likely diagnosis. This is efficient when it works. It is lethal when it doesn't.

Then there is **premature closure**—the tendency to stop considering alternative diagnoses once a satisfactory one has been found. In the language of cognitive science, the physician "closes" the diagnostic process prematurely, accepting the first explanation that fits the available data without stress-testing it against competing hypotheses. Premature closure is especially dangerous in emergency medicine, where time pressure and patient volume create an almost irresistible incentive to reach a conclusion quickly and move on to the next case. The ER physician seeing forty-three patients in a shift does not have the luxury of dwelling on a single case. The system demands speed, and speed demands shortcuts, and shortcuts demand trust in first impressions. Sometimes that trust is well placed. Sometimes a man dies of a heart attack that was mistaken for heartburn.

The statistics are sobering. Research published in the journal BMJ Quality & Safety estimates that diagnostic errors affect approximately twelve million adults in the United States every year in outpatient settings alone. Of these, roughly half have the potential to result in serious harm. In emergency departments, the misdiagnosis rate for certain time-sensitive conditions—stroke, heart attack, pulmonary embolism—can reach as high as twenty to thirty percent, depending on the study and the condition. These are not rare diseases being missed by overwhelmed residents. These are common, well-understood, eminently treatable conditions that experienced clinicians are missing because the cognitive architecture of the human brain is not built to meet the demands that modern medicine places on it.

The tragedy is compounded by the fact that the solution is known. Structured diagnostic checklists, mandatory "diagnostic time-outs" in which the clinician pauses to consider alternative explanations, and decision-support software that prompts physicians to consider red-flag diagnoses have all been shown to reduce diagnostic error rates significantly in clinical trials. Yet adoption of these tools remains stubbornly low across the industry. The reasons are cultural as much as logistical. Medicine has a long tradition of prizing individual clinical judgment—the idea that the experienced physician, through intuition and pattern recognition, can arrive at the correct diagnosis faster and more reliably than any checklist or algorithm. This reverence for individual expertise is not entirely unfounded; clinical intuition is real and often remarkable. But it is also fallible, and in a system where fallibility kills, reverence for intuition must yield to evidence.

There is one more cognitive trap that deserves particular attention, because it operates not before the diagnosis is made, but after: confirmation bias. Once a physician has settled on a working diagnosis, the mind begins to selectively seek out information that supports it and to minimize or dismiss information that contradicts it. A lab result that is slightly abnormal but does not fit the working diagnosis is rationalized away as a lab error. A nurse's observation that the patient "just doesn't look right" is filed under subjective impression rather than clinical data. A family member's insistence that something has changed is attributed to anxiety rather than observation. Confirmation bias does not merely prevent the physician from seeing the truth. It actively constructs a false reality in which the wrong diagnosis appears increasingly well-supported, even as the patient's actual condition deteriorates.

Taken together, these biases form a cognitive trap from which even the most brilliant diagnostician can struggle to escape. The anchoring bias sets the initial frame. Availability bias reinforces it with the false authority of recent experience. Premature closure locks it in place. And confirmation bias builds walls around it, filtering out any evidence that might prompt a second look. The diagnosis is not just wrong; it is defended—unconsciously, reflexively, and with increasing conviction—right up to the moment when the patient codes and the truth arrives too late to matter.

Communication Breakdown: The "Silo Effect"

If diagnostic failure is a disease of the individual mind, communication breakdown is a disease of the institutional body. It occurs when critical information—a lab result, a change in patient status, a medication allergy, a verbal instruction—fails to travel from the person who has it to the person who needs it. The failure can happen at any point in the chain: during a shift handoff between nurses, during a transfer between departments, during a phone call between a specialist and a primary care physician, or during the deceptively simple act of entering data into an electronic health record.

The healthcare industry calls this the "Silo Effect"—a term borrowed from organizational theory to describe the tendency of departments and teams to operate as self-contained units, each with its own information systems, communication protocols, and institutional culture. The emergency department operates in one silo. The radiology department operates in another building. The laboratory operates in a third. The pharmacy operates in a fourth. Each silo is internally functional. The problem is at the seams—the places where one silo must hand information to another, often through shockingly primitive mechanisms for an industry that prides itself on technological sophistication.

Consider a scenario that plays out thousands of times every day in hospitals across the country. A patient in the emergency department has blood drawn for a panel of tests. The blood is sent to the laboratory. The results come back flagging a critically elevated Troponin level—a marker that, in the context of the patient's symptoms, strongly suggests an active heart attack. The lab technician sees the result and enters it into the electronic health record system. The result is now, technically, "available." It exists in the database. It is accessible to any physician with the proper credentials and the inclination to look.

But "available" and "seen" are not the same thing. The attending physician in the ER is currently managing three critical patients simultaneously. Her electronic inbox contains forty-seven unread notifications. The Troponin result does not arrive with a siren or a red flag on her screen. It arrives as one more line item in a cascade of data, sandwiched between a routine CBC and a urinalysis. If she is not actively looking for it—if, for instance, she has already anchored on a provisional diagnosis and moved on to the next patient—the result will sit unread. The patient may be discharged. The heart attack may happen at home.

This is not a hypothetical. This is Elena's story, which we will explore in full in the next chapter. And Elena's story is not an anomaly. Research conducted by the Agency for Healthcare Research and Quality found that communication failures are the leading root cause of sentinel events reported to the Joint Commission—more common than clinical judgment errors, more common than technical failures, more common than any other single factor. An estimated eighty percent of serious medical errors involve miscommunication during patient handoffs.

The handoff—the moment when responsibility for a patient transfers from one provider to another—is perhaps the most dangerous ritual in all of medicine. It typically occurs during shift changes, which in most hospitals happen at seven o'clock in the morning and seven o'clock in the evening. In the space of a few minutes, the outgoing nurse or physician must compress hours of observation, dozens of data points, and a constellation of clinical impressions into a verbal summary that the incoming provider can absorb while simultaneously fielding pages, checking new orders, and mentally preparing for the shift ahead.

The opportunities for loss are staggering. A detail mentioned in passing may not be heard. A nuance of tone—"she seemed a little off to me"—may not be transmitted, because clinical suspicion without objective data is difficult to articulate and easy to dismiss. An order that was verbal but never entered into the system may be forgotten entirely. The incoming provider inherits not the complete picture of the patient, but a compressed, filtered, and inevitably degraded version of it. What arrives on the other side of the handoff is not the patient. It is a shadow of the patient—and sometimes, the details lost in that shadow are the ones that matter most.

The SBAR technique—Situation, Background, Assessment, Recommendation—was developed specifically to address this problem. Borrowed from the United States Navy's submarine force, where miscommunication can result in catastrophe in enclosed and unforgiving environments, SBAR imposes a standardized structure on the handoff process. It requires the outgoing provider to organize information into four distinct categories, ensuring that no critical element is omitted. Hospitals that have adopted SBAR protocols have seen measurable reductions in handoff-related errors. Yet remarkably, many hospitals still have no standardized handoff protocol, leaving one of the most dangerous moments in patient care to the vagaries of individual habits, memory, and fatigue.

Verbal orders compound the danger. In the controlled chaos of a busy ward or an emergency resuscitation, physicians frequently issue instructions aloud rather than entering them into the electronic system. "Give 5 milligrams of Morphine IV." "Start a Heparin drip at 18 units per kilo per hour." These orders are spoken once, often quickly, often in noisy environments, and often to a nurse who is simultaneously performing another task. The margin for mishearing is enormous. "Fifteen" and "fifty" sound nearly identical across a crowded room. "Milligrams" and "micrograms" are easily confused. Without a read-back protocol—a requirement that the nurse repeat the order aloud and the physician confirm it—a single misheard syllable can deliver a tenfold overdose or a critically insufficient dosage. The fix is simple and costs nothing. Yet in many institutions, read-back remains a guideline rather than a mandate, observed when convenient and abandoned when the pace of care accelerates—which is, of course, precisely when the risk of miscommunication is highest.

There is also a hierarchy problem. Medicine is, by tradition and by training, a deeply hierarchical profession. The attending physician sits at the top. Residents, nurses, technicians, and support staff descend in order of perceived authority. This hierarchy has legitimate functions—someone must bear ultimate responsibility for clinical decisions. But it also creates a communication gradient that is lethally counterproductive for patient safety. A nurse who notices something worrying but is reluctant to challenge an attending physician's judgment. A medical student who spots an error on a medication order but hesitates to speak up because the prescribing physician is a department chief. A respiratory therapist who thinks the ventilator settings are wrong but defers to the intensivist. In each case, the hierarchy does not merely permit silence. It incentivizes it. And in a system where silence kills, any incentive for silence is an incentive for harm.

Technological Over-Reliance: When We Trust the Monitor More Than the Patient

The modern hospital is, by any measure, a technological marvel. Electronic health records have replaced paper charts. Computerized physician order entry has replaced handwritten prescriptions. Smart infusion pumps calculate dosages automatically. Pulse oximeters, cardiac monitors, and automated blood pressure cuffs provide continuous streams of physiological data, painting a real-time portrait of the patient's condition that would have seemed miraculous to physicians of even a generation ago.

And yet, for all its sophistication, this technology has introduced a new and insidious category of risk: the danger of trusting the machine more than the human being it is monitoring. This is not a Luddite argument against technology. The tools of modern medicine have saved countless lives and will continue to do so. But technology is not neutral. It shapes behavior. It redirects attention. And when it fails—as all technology eventually does—the consequences can be catastrophic, precisely because the humans in the system have been trained to defer to it.

The concept at work here has a name in human factors engineering: automation complacency. It describes the well-documented tendency of human operators to become less vigilant when they believe a machine is monitoring a situation on their behalf. Studies in aviation—an industry that confronted this problem decades before healthcare began to acknowledge it—have shown that pilots monitoring automated flight systems are significantly slower to detect anomalies than pilots flying manually, even when the anomalies are blatant. The autopilot does not make the pilot less skilled. It makes the pilot less attentive. The same dynamic plays out in hospital rooms every day.

A nurse caring for a postoperative patient checks the pulse oximeter and notes an oxygen saturation of 97%—normal. She records the number, checks the intravenous line, and moves on to her next patient. What she does not do is look at the patient's face. She does not notice the subtle grayish tinge around his lips. She does not observe that his breathing, while still technically within the monitor's acceptable range, has become shallower and more effortful over the past hour. The monitor says he is fine. The monitor is a machine that measures one variable at a time, through a sensor clipped to a fingertip. The human body is not one variable. It is a symphony of interdependent systems, and the earliest signs of distress are often not quantitative but qualitative—a change in color, a change in posture, a change in the quality of a breath that no sensor can capture.

This is David's story, revisited through the lens of technology. The PCA pump that delivered his pain medication was functioning exactly as designed. The pulse oximeter that monitored his oxygen was functioning exactly as designed. The alarm that triggered when his saturation dropped was functioning exactly as designed. Every piece of technology in the room performed its intended function. And David still died, because the technology had become a substitute for, rather than a supplement to, human observation. The staff had been trained to trust the monitors. The monitors, in turn, had been configured with thresholds so broad that they generated hundreds of alarms per shift, most of them clinically insignificant. The technology did not fail. The system of human-technology interaction failed.

There is a deeper irony here. Electronic health record systems, which were introduced in part to reduce errors caused by illegible handwriting, misplaced charts, and fragmented documentation, have created an entirely new ecosystem of risk. Physicians now spend, by some estimates, nearly half of their working hours interacting with EHR systems—entering data, navigating menus, clicking through alerts, copying and pasting progress notes from one day to the next. This is time that is not spent at the bedside. It is time that is not spent talking to the patient, examining the patient, or simply observing the patient with the kind of sustained, unhurried attention that, despite all technological advances, remains the most sensitive diagnostic instrument in medicine: the trained human eye.

The phenomenon has become so widespread that it has earned its own clinical term: "pajama time"—the hours that physicians spend at home after their shifts end, completing electronic documentation they could not finish during the workday because the workday was consumed by electronic documentation. It is a vicious cycle that feeds directly into physician burnout, cognitive fatigue, and, inevitably, the kind of attentional lapses that lead to missed diagnoses, overlooked lab results, and unheard alarms.

The copy-and-paste function within EHR systems introduces yet another layer of risk. To save time, physicians routinely copy a previous day's progress note and paste it into the current day's record, modifying only the details that have changed. The practice is understandable—rewriting an entire note from scratch for a patient whose condition has not dramatically shifted is tedious and, frankly, unnecessary. But the danger lies in what is not updated. A note from Tuesday that says "patient alert and oriented, no respiratory distress" may be pasted into Wednesday's record without modification, even if the patient's condition has subtly shifted overnight. The medical record now contains a lie—not a deliberate one, but a lie of omission baked into the software's efficiency. Future clinicians reading the chart will see a continuous narrative of stability that does not reflect reality. Decisions will be made based on documentation that has been copied, not observed. The chart becomes a palimpsest of outdated assessments, each layer obscuring the one beneath it.

Then there is alert fatigue within the EHR itself—a digital cousin of the alarm fatigue that plagues bedside monitors. Modern electronic health record systems are designed to generate clinical alerts: warnings about potential drug interactions, allergy notifications, dosage range violations, and duplicate order flags. In theory, these alerts are a safety net. In practice, they are a deluge. Studies have shown that physicians are confronted with an average of fifty to seventy clinical alerts per day, the overwhelming majority of which are overridden without review. The override rate for drug interaction alerts in some systems exceeds ninety percent. When nine out of ten warnings are clinically irrelevant, the tenth—the one that matters, the one that could prevent a lethal drug combination or catch a contraindicated prescription—is buried under the accumulated weight of all the ones that did not.

The solution is not to abandon technology. That ship has sailed, and there is no going back—nor should there be. The solution is to redesign the relationship between human beings and the machines they use. Smart alarm systems that filter out noise and escalate only genuine emergencies. EHR interfaces are designed around clinical workflow rather than billing requirements. Decision-support tools that augment clinical judgment without replacing it. Monitoring systems that integrate multiple data streams—heart rate, respiratory rate, blood pressure, oxygen saturation—into a single, intelligible picture of the patient's trajectory, rather than presenting each variable in isolation on separate screens. The technology exists. What is lacking is the institutional will to deploy it in a way that serves the patient rather than the spreadsheet.

The Common Thread

Diagnostic failure, communication breakdown, and technological over-reliance are distinct categories, but they do not operate in isolation. In practice, they interact, overlap, and amplify one another in ways that make the resulting errors far more dangerous than any single failure could be on its own.

A physician anchors on the wrong diagnosis because she is cognitively fatigued—cognitive fatigue that was caused, in part, by three hours of EHR documentation before she ever saw the patient. The critical lab result that might have corrected her course is entered into the system but never reaches her, because the EHR's notification algorithm buried it among dozens of routine alerts. The nurse who might have noticed the patient's deterioration was monitoring six patients instead of four because the hospital's staffing model had been optimized for cost efficiency, and she had been trained to trust the pulse oximeter over her own instincts. By the time the alarm sounds, it is one of three hundred that shift, and no one hears it.

This is the anatomy of a medical error. Not a single point of failure, but a web of interconnected vulnerabilities—cognitive, communicative, and technological—woven so tightly into the fabric of daily hospital operations that they become invisible to the people working inside them. The error is not an aberration. It is the system functioning exactly as it was built to, with the exact consequences its architecture makes inevitable.

In the next chapter, we will see this anatomy in action. Through the stories of David and Elena—two patients whose deaths were as preventable as they were predictable—we will trace each strand of the web, from the first missed signal to the final, irreversible outcome. Their stories are not unique. They are, in the deepest and most troubling sense, ordinary. And that is precisely why they demand to be told.

CHAPTER THREE

When the System Fails

Dr. Dorathy U. Ubiam

> *"Every system is perfectly designed
> to get the results it gets."*
>
> — W. Edwards Deming

In the previous chapter, we examined the anatomy of medical error in the abstract—the cognitive biases that distort diagnosis, the communication failures that fracture information, the technological dependencies that dull human vigilance. These are the structural vulnerabilities of the system, the fault lines that run beneath the surface of every hospital in the country. But fault lines are invisible until they rupture. This chapter is about the rupture.

What follows are two case studies. They are drawn from the kinds of incidents that occur in hospitals every day, reconstructed in narrative detail to illuminate not just what happened, but why it happened—and, crucially, how it could have been prevented. Each case follows a deliberate structure: the patient, the incident, the outcome, and the mitigation lesson. This structure is not arbitrary. It reflects the forensic methodology used in root cause analysis, the investigative framework that hospitals are supposed to employ after every serious adverse event. The difference is that root cause analyses are typically conducted behind closed doors, written in clinical language, and filed in institutional archives where they gather dust. These stories are told in the open, in human language, because the lessons they contain are too important to remain locked in committee reports.

A word of preparation. These stories are difficult to read. They involve real suffering, real loss, and real institutional failure. That difficulty is the point. If we sanitize these narratives, if we strip them of their emotional weight in pursuit of clinical objectivity, we strip them of the very quality that makes them powerful enough to change minds and, ultimately, to change systems.

Case Study 1: The Echo of Silence

The 2:00 AM Alarm That Nobody Heard

The Patient

David Morales was forty-two years old. He worked as a project manager for a mid-sized construction firm and had spent the past eighteen years building a life that revolved around three things: his work, his family, and the under-twelve girls' soccer team he coached every Saturday morning at the municipal park. His wife, Maria, was a high school Spanish teacher. Their daughter, Sofia, was eleven—old enough to understand that her father's knee surgery was routine, young enough to make a get-well card with a lopsided soccer ball on the front and the words "Score a goal for me, Dad!" written in purple marker.

David had torn his anterior cruciate ligament during a company softball game—a middle-aged injury so common it had become almost a rite of passage among his colleagues. His orthopedic surgeon, Dr. Rajesh Patel, had performed the procedure hundreds of times. The surgery was scheduled for a Tuesday morning, and David was expected to be home by Thursday. He packed a bag with loose sweatpants and a paperback novel. Maria kissed him at the hospital entrance and told him she would pick him up in two days. Neither of them considered, even for a moment, that this was the last conversation they would have.

The Incident

The arthroscopic repair went smoothly. David was moved to a post-operative recovery room on the fourth floor at approximately 3:00 PM. By 6:00 PM, he was alert, in moderate pain, and connected to a Patient-Controlled Analgesia pump—a device that allows patients to self-administer small, pre-programmed doses of intravenous opioids by pressing a button. The PCA pump was loaded with hydromorphone, a powerful synthetic opioid approximately five times more potent than morphine by weight. The pump was programmed with a lockout interval to prevent overdose, and a pulse oximeter was clipped to David's right index finger to monitor his blood oxygen saturation continuously.

David used the PCA pump several times during the evening. His pain was managed. His oxygen saturation remained in the normal range. He spoke to Maria by phone at 9:30 PM and told her the worst part was the hospital food. She laughed. He told her he loved her. At 10:00 PM, the night shift arrived.

The night shift that evening was staffed at a ratio of one nurse to six patients—a ratio that the hospital's own internal safety guidelines identified as the maximum acceptable threshold for a post-surgical unit. The nurse assigned to David's section, a twelve-year veteran named Karen Liu, was working the second of three consecutive twelve-hour overnight shifts. She had arrived at 7:00 PM and would not leave until 7:30 AM. By midnight, she had already been on her feet for seventeen hours that day, factoring in errands and childcare before her shift began.

At approximately 11:45 PM, David's respiratory rate began to decline. Opioid-induced respiratory depression is a well-known and predictable complication of PCA therapy—the very reason the pulse oximeter had been attached. As David's breathing slowed, his blood oxygen saturation began a gradual descent: from ninety-seven percent at 11:00 PM, to ninety-four percent at midnight, to ninety-one at 12:15 AM. At 12:22 AM, the pulse oximeter registered eighty-eight percent, and the bedside alarm activated.

The alarm sounded in the hallway outside David's room. It was a tone that Karen Liu and her colleagues heard, by conservative estimate, between 200 and 350 times per twelve-hour shift. The vast majority of these alarms were clinically insignificant—a sensor displaced by a patient rolling over in bed, a momentary dip caused by a clenched fist, an artifact generated by patient movement during sleep. Studies have consistently shown that between 85 and 99 percent of bedside monitor alarms in hospital settings are false or non-actionable. The term for the cumulative effect of this relentless electronic noise is alarm fatigue, and it is not a metaphor. It is a physiologically measurable phenomenon: after sustained exposure to frequent alarms, the human brain literally attenuates its auditory response, processing the sound at a lower cognitive priority. The alarm does not become inaudible. It becomes insignificant—background noise, like traffic or air conditioning.

Karen Liu heard the alarm at 12:22 AM. She was in the middle of administering medication to another patient two rooms away. She noted the alarm, assumed it was a sensor displacement—as it had been the previous four times that evening—and planned to check on David as soon as she finished her current task. She did not finish her current task for another eleven minutes, because the patient she was attending required an unscheduled IV line change. By the time she reached David's room at 12:33 AM, his oxygen saturation had dropped to eighty-two percent. She repositioned the sensor and noted that he appeared to be sleeping deeply. She did not manually assess his respiratory rate. She did not perform a sternal rub or other stimulation to assess his level of consciousness. She adjusted the sensor, saw the reading climb back to eighty-nine percent, and returned to her rounds.

Over the next ninety minutes, David's condition continued to deteriorate. The alarm sounded twice more—at 1:08 AM and 1:47 AM. Each time, it was acknowledged remotely by a nurse at the central monitoring station, who silenced it and logged the reading without dispatching anyone to the bedside. The system allowed this. The protocol permitted it. At no point did the alarm escalate. At no point did the PCA pump receive a signal to pause or reduce the opioid infusion. The technologies in the room—the pump, the oximeter, the alarm—operated in perfect isolation from one another, each performing its individual function without any awareness of the others.

At 2:07 AM, David Morales stopped breathing. His heart, starved of oxygen for an indeterminate period, entered a fatal arrhythmia. The pulse oximeter registered zero. The alarm that sounded this time was different—a sustained, unbroken tone that indicated not a dip, but a flatline. A code blue was called at 2:15 AM. The resuscitation team arrived within four minutes. They worked on David for thirty-seven minutes. He was pronounced dead at 2:56 AM.

The Aftermath

Maria Morales received the phone call at 3:47 AM. She drove to the hospital in her pajamas. When she arrived, a physician she had never met explained, in the measured cadence of someone who has delivered this news before, that David had suffered a "respiratory event" and that despite the team's best efforts, he could not be resuscitated. Maria asked what had happened. The physician said it was a known complication of post-operative pain management. He did not mention the alarms. He did not mention alarm fatigue. He did not mention that the pulse oximeter and the PCA pump were not linked—that the machine delivering the drug that suppressed David's breathing had no connection to the machine measuring the consequences.

Sofia's get-well card was still on the kitchen counter. The loose sweatpants were still folded in a bag by the front door. Maria would later tell a reporter that she kept the bag there for three weeks because moving it felt like admitting that David was never coming home, and she was not ready to admit that.

The hospital's internal review classified David's death as a "contributing factor: delayed response to clinical alarm." Karen Liu was counseled but not disciplined. The review committee noted that staffing ratios were "within acceptable parameters" and that the alarm system was "functioning as designed." Both of these statements were technically true. Both of them were the problem.

Karen Liu, for her part, became a second victim. In the weeks following David's death, she replayed the night endlessly—the moment she heard the alarm at 12:22 AM, the eleven minutes it took her to reach his room, the decision to reposition the sensor instead of assessing his breathing manually. She had made a judgment call based on 12 years of experience, during which the overwhelming majority of alarms she had responded to turned out to be false alarms. Her brain had done exactly what it had been conditioned to do by a system that flooded her with meaningless alerts until the meaningful ones became indistinguishable. She requested a transfer to an outpatient clinic three months later. She told a colleague that she could no longer hear a monitor alarm without her hands shaking.

The Mitigation Lesson

David's death was preventable at multiple points, through interventions that already exist and have been proven effective. The first and most critical is **vitals-linking technology**—the integration of the PCA pump with the pulse oximeter so that when oxygen saturation drops below a predetermined threshold, the pump automatically suspends opioid delivery. This technology exists. It is commercially available. It has been endorsed by patient safety organizations, including the Anesthesia Patient Safety Foundation. Yet at the time of David's death, his hospital had not implemented it, because the capital expenditure required to retrofit existing PCA pumps had been deferred in three consecutive budget cycles.

The second intervention is **smart alarm management**—the replacement of one-size-fits-all alarm thresholds with patient-specific baselines, which dramatically reduces false-positive rates. When an alarm sounds only when something genuinely abnormal is occurring, clinicians respond. When it sounds 300 times per shift for no clinically significant reason, they stop responding. This is not a character failing. It is a mathematical certainty. The alarm system that killed David was designed to detect every possible deviation. In doing so, it detected nothing.

The third intervention is **acuity-based staffing**—matching nurse-to-patient ratios not to a fixed institutional standard, but to the actual clinical complexity of the patients on the unit. A post-operative patient on continuous opioid infusion is not the same as a patient recovering from an uncomplicated appendectomy. Treating them as identical for staffing purposes is not efficient. It is negligence with a spreadsheet.

Case Study 2: The Paper Trail to Nowhere

The Missing Lab and the Lost Mother

The Patient

Elena Vasquez was sixty-five years old, a retired postal worker, and a grandmother of four. She lived alone in a small apartment three miles from the hospital, having lost her husband, Tomás, to pancreatic cancer four years earlier. Her children—two daughters and a son—lived within driving distance and visited regularly, though not as often as Elena would have liked, and more often than they realized she needed. She was a woman of routines: morning coffee with the newspaper, an afternoon walk to the community garden where she tended a plot of tomatoes and marigolds, and a nightly phone call with her eldest daughter, Carmen, that never lasted less than forty-five minutes.

On a Thursday evening in late September, Elena began to feel a dull tightness in her chest. She attributed it to the enchiladas she had eaten for dinner—too much cheese, she told herself. By 10:00 PM, the discomfort had not subsided, and a faint nausea had settled in her stomach. Elena was not a woman who visited emergency rooms lightly. She had grown up in a household where illness was endured, not indulged. But something about the persistence of the sensation unsettled her, and at 10:30 PM, she called Carmen, who insisted on driving her to the hospital immediately.

The Incident

Elena arrived at the emergency department at 11:15 PM. She was triaged, assigned a bed, and seen by the attending physician, Dr. Michael Torres, within forty minutes—a relatively efficient timeline for a busy urban ER on a Thursday night. She described the chest tightness to Dr. Torres, who noted her age, her vague symptom presentation, and the absence of classic cardiac risk factors beyond her age and her gender. An electrocardiogram was ordered. It came back normal. A chest X-ray was ordered. It was unremarkable. Blood was drawn for a standard cardiac panel, including a Troponin-I assay—the gold standard biomarker for detecting myocardial injury.

This is the moment where the system began to fail, though no one in the room knew it. The blood sample was sent to the hospital's central laboratory at approximately 12:30 AM. The lab processed the sample and generated the result by 1:15 AM. Elena's Troponin-I level was 2.4 nanograms per milliliter— more than twenty times the upper limit of normal. In the context of her symptoms, this result was an unambiguous red flag: Elena was almost certainly in the early stages of a myocardial infarction. The result demanded immediate clinical intervention—at minimum, a repeat Troponin, serial ECGs, anticoagulation therapy, and a cardiology consultation.

The result never reached Dr. Torres. At 1:00 AM—fifteen minutes before the Troponin result was finalized—the ER had undergone its nightly shift change. Dr. Torres had handed off his patients to the incoming attending, Dr. Sarah Kim, using a verbal summary at the nursing station. In that summary, he described Elena as a "sixty-five-year-old female with atypical chest pain, ECG normal, chest X-ray clear, cardiac enzymes pending." The critical word was "pending." It signaled that the result had not yet arrived and that someone would need to follow up. But in the fluid architecture of a shift change—where dozens of patients, hundreds of data points, and an entire night's worth of clinical context must be compressed into a few minutes of conversation—"pending" did not carry the weight it needed to carry. It was one detail among many, and it slipped.

Compounding the handoff failure was a software malfunction. The hospital's electronic health record system was designed to push critical lab values to the ordering physician's inbox as a high-priority notification. But the system identified the ordering physician as Dr. Torres, who had logged out at the end of his shift. The notification was delivered to his inbox, where it sat unread. Dr. Kim, who was now responsible for Elena's care, did not receive the notification because the system's architecture assigned results to the ordering physician rather than the covering physician. There was no automated escalation. There was no secondary alert. There was no fail-safe. The critical lab value existed in the database, technically accessible to anyone who thought to look for it, but actively delivered to no one who could act on it.

At 2:45 AM, Dr. Kim reviewed Elena's chart. She saw a normal ECG, a clear chest X-ray, and a notation that cardiac enzymes were pending. She did not check the lab system for results, because the EHR's dashboard showed no outstanding critical alerts for Elena's file. The absence of an alert was reasonably but fatally interpreted as the absence of a problem. At 3:20 AM, Elena was discharged with a diagnosis of gastroesophageal reflux disease and a prescription for an antacid. Carmen drove her mother home.

Elena went to bed at approximately 4:00 AM. She told Carmen she was feeling a little better. Carmen kissed her mother on the forehead and drove home, planning to check on her in the morning. At approximately 9:30 AM, when Carmen called and received no answer, she drove to the apartment. She found Elena in her bed, still in the clothes she had worn home from the hospital. She was not breathing. The paramedics pronounced her dead at the scene. The cause of death, determined by autopsy, was a massive ST-elevation myocardial infarction—the final, catastrophic consequence of the heart attack that had been quietly progressing since Thursday evening, the one that a single lab result, delivered to the right person at the right time, would have detected and almost certainly prevented.

The Aftermath

Carmen Vasquez would spend the next six months trying to understand how her mother's death was possible. How a woman could walk into an emergency room with chest pain, have a blood test that confirmed she was having a heart attack, and be sent home to die of that heart attack. The hospital's initial response was silence. When Carmen filed a formal complaint, she received a letter acknowledging her "concerns" and assuring her that the case would be reviewed by the quality improvement committee. She was not invited to participate in the review. She was not told what the review found. She learned about the Troponin result—the result that had been available for more than two hours before her mother was discharged—only after retaining an attorney who subpoenaed the medical records.

What haunted Carmen most was the timeline. She could reconstruct it now, with the medical records in front of her: at 1:15 AM, the lab had finalized a result that unambiguously identified a heart attack in progress. At 3:20 AM—more than two hours later—her mother had been discharged with antacids. During those two hours, the information that could have saved Elena's life existed in the hospital's database, accessible to anyone with a login and a reason to look. No one looked. The system had generated the data. The system had failed to deliver it. And Elena had been sent home to die in the apartment where she had lived alone since Tomás passed—found the next morning in the same clothes she had worn to the emergency room, as though she had been too tired, or too trusting, to change before lying down for what she believed would be a normal night's sleep.

The internal review identified two root causes: the EHR notification routing failure and the absence of a standardized lab result verification protocol during shift changes. Recommendations were issued. A software patch was requested. A policy memo was distributed. Dr. Torres and Dr. Kim were neither disciplined nor named in any public document. The hospital's official position was that the error was systemic, not individual—a position that was, in this case, entirely correct and entirely inadequate, because being correct about the nature of the failure means nothing if the system that produced it remains fundamentally unchanged.

The Mitigation Lesson

Elena's death points to two mitigation strategies that, had they been in place, would almost certainly have saved her life. The first is **closed-loop reporting**—a system design principle that requires every critical lab value to be acknowledged by a responsible clinician before any disposition change (admission, discharge, or transfer) can be processed in the electronic record. Under a closed-loop protocol, the EHR would not have permitted Elena's discharge order to be finalized until a physician had electronically acknowledged receipt of the Troponin result. The system would have flagged the unacknowledged critical value and physically blocked the discharge workflow. This is not a theoretical concept. Closed-loop systems are standard in aviation, nuclear power, and pharmaceutical manufacturing.

In healthcare, their adoption remains inconsistent and, in many institutions, nonexistent.

The second is the **"Second Look" Protocol**—a mandatory review of all diagnostic data by a discharge nurse or physician assistant before any patient leaves the emergency department. The Second Look is not a duplication of effort. It is a redundancy of safety—a second pair of eyes scanning for precisely the kind of result that fell through the cracks in Elena's case. It adds approximately four to seven minutes to the discharge process. Elena's life was worth more than seven minutes.

The Pattern Beneath the Stories

David and Elena never met. They were treated in different hospitals, by different teams, for entirely different conditions. And yet their stories share a structural DNA that is impossible to ignore. In both cases, the technology functioned as designed and failed the patient anyway. In both cases, the staffing model prioritized efficiency over vigilance. In both cases, a critical piece of information—an alarm, a lab result— was available but never acted upon. And in both cases, the system's post-mortem response was to identify the failure, issue a recommendation, and move on, leaving the fundamental architecture intact.

This is the pattern. It is not a coincidence. It is the predictable, repeatable output of a healthcare system optimized for throughput and liability management rather than for patient safety. David's alarm was heard and ignored because the system generated so many false alarms that the real one was indistinguishable from the noise. Elena's lab result was available and unseen because the system routed critical information to an inbox that no one was monitoring. In both cases, the error was not a deviation from the system. It was the system operating exactly as built.

Perhaps most troubling is what happened after each death. Both hospitals conducted internal reviews. Both identified root causes. Both issued recommendations. And both, in all probability, returned to business as usual within weeks. This is the final and most devastating failure in the chain: the failure to learn. Root cause analyses, when they are conducted at all, produce reports that are read by committee members and filed in institutional archives. The recommendations they generate are often vague: "improve communication during handoffs," "consider alarm management strategies," and are rarely accompanied by timelines, budgets, or mechanisms for accountability. A recommendation without implementation is not a safety measure. It is a legal document. It exists to demonstrate that the hospital "took action" in the event of litigation. It does not exist to prevent the next David or the next Elena. And until the incentive structure changes—until hospitals are rewarded for preventing harm as aggressively as they are penalized for causing it—the paper trail of recommendations will continue to lead exactly where Elena's lab result led: nowhere.

The chapters that follow will propose a different architecture—one designed not to assign blame after the fact, but to prevent harm before it occurs. The mitigation frameworks we will explore are not hypothetical. They are proven, practical, and implementable. What they require is not technological innovation or medical genius, but something far more difficult to summon: the institutional courage to admit that the current system is failing, and the collective will to build something better.

David and Elena deserve that. So do the patients who will occupy their beds tomorrow.

CHAPTER FOUR

The Mitigation Framework

Dr. Dorathy U. Ubiam

"You cannot inspect quality into a product. You have to build it in."

— W. Edwards Deming

The previous chapters have cataloged a crisis. We have traced the anatomy of medical error through its cognitive, communicative, and technological dimensions. We have followed David Morales and Elena Vasquez from hospital admission to preventable death, reconstructing in forensic detail the cascading failures that took their lives. We now arrive at the question that justifies this entire book: What do we do about it?

The answer is not a single solution. There is no silver bullet for a problem this deeply embedded in the fabric of healthcare delivery. What exists instead is a framework—a layered, interlocking set of strategies drawn from industries that have already confronted the challenge of reducing preventable human death to near zero. Aviation. Nuclear power. Automotive manufacturing. These industries did not achieve their safety records through goodwill or individual heroism. They achieved them through systemic redesign—by building safety into the architecture of their operations so thoroughly that catastrophic failure requires not one lapse, but an almost impossible alignment of multiple simultaneous breakdowns.

This chapter presents four pillars of that framework. Each addresses a distinct category of failure identified in the preceding chapters. Together, they form a comprehensive mitigation strategy that, if implemented with institutional commitment, could transform patient safety from an aspiration into an operational reality.

1. The "Swiss Cheese" Model of Safety

In 1990, a British psychologist named James Reason published a paper that would fundamentally alter the way safety scientists think about accidents. Reason proposed that most accidents in complex systems are not caused by a single catastrophic failure, but by the alignment of multiple smaller failures—each one, taken individually, survivable. He illustrated his theory with an image that has since become iconic in patient safety literature: slices of Swiss cheese stacked side by side.

Each slice represents a layer of defense in a system—a safety protocol, a verification step, a redundancy check. Each slice has holes—imperfections, gaps, vulnerabilities inherent in any human-designed process. In a well-designed system, the holes in one slice are covered by the solid portions of the adjacent slice. No single layer is expected to be perfect. The system achieves safety not through the perfection of any individual component, but through the cumulative effect of multiple imperfect layers arranged so that their vulnerabilities do not align. An accident occurs only when the holes in every slice line up simultaneously, creating a clear trajectory from hazard to harm—what Reason called an "accident trajectory."

The implications for healthcare are immediate and profound. Consider David Morales. The first slice of cheese—the PCA pump's built-in lockout interval—had a hole: it prevented overdose from rapid button-pressing but did not account for the slow, cumulative respiratory depression that occurs over hours. The second slice—the pulse oximeter alarm—had a hole: it triggered correctly but was indistinguishable from hundreds of false alarms. The third slice—the bedside nurse—had a hole: she was cognitively depleted from consecutive overnight shifts and caring for six patients simultaneously. The fourth slice—the central monitoring station—had a hole: it permitted remote acknowledgment without bedside verification. Each of these layers, in isolation, was a reasonable defense. But every one of them had a gap, and on the night David died, those gaps aligned.

The Swiss Cheese Model teaches us that the solution is not to demand perfection from any single layer. It is to add more layers and to ensure that their holes do not overlap. In practical terms, this means that every high-risk procedure or clinical scenario must have at least three independent checkpoints—safeguards designed by different teams, relying on different mechanisms, and governed by different failure modes. A surgical safety checklist, for instance, requires the surgeon, nurse, and anesthesiologist to independently verify the patient's identity, surgical site, and planned procedure before the first incision. The surgeon's verification covers for the nurse's potential error. The anesthesiologist's verification covers for the surgeon's potential distraction. No one person bears the full weight of safety. The system does.

The World Health Organization's Surgical Safety Checklist, introduced in 2008, demonstrated the power of this approach in dramatic fashion. A study published in the New England Journal of Medicine found that hospitals implementing the checklist saw a 36% reduction in major surgical complications and a 47% reduction in mortality. These results were achieved not through new technology or additional staffing, but through a simple, one-page checklist that imposed three independent verification points on a process that had previously relied on the assumption that everyone in the room already knew what was happening. The Swiss Cheese Model does not assume competence is sufficient. It assumes competence is fallible, and it builds accordingly.

The model extends well beyond the operating room. In medication administration, a properly designed Swiss Cheese system includes at least five layers: the prescribing physician selects the drug and dose; the computerized physician order entry system checks for allergies, interactions, and dosage range violations; the pharmacist reviews and verifies the order; the dispensing system confirms the medication identity via barcode; and the bedside nurse performs a final check against the patient's wristband before administration. Each layer has vulnerabilities. The physician may select the wrong drug from a drop-down menu. The software may fail to flag an interaction not yet in its database. The pharmacist may be processing a hundred orders simultaneously. The barcode scanner may malfunction. The nurse may be interrupted mid-verification. But for a medication error to reach the patient, every single one of these layers must fail simultaneously. The probability of that alignment is vanishingly small—orders of magnitude smaller than the probability of any single layer failing on its own.

For the discharge process—the process that killed Elena Vasquez—the Swiss Cheese Model demands at minimum three independent checks before a patient leaves the hospital: the attending physician's clinical assessment, an electronic verification that all pending results have been reviewed and acknowledged, and a discharge nurse's independent review of the complete diagnostic record. Elena's hospital had one layer: the attending physician's judgment. One slice of cheese. One hole. One death.

2. Eradicating Alarm Fatigue

David Morales died because an alarm sounded and no one came. Understanding why requires us to confront one of the most perverse paradoxes in modern hospital design: the tools built to keep patients safe are, through their sheer volume, making patients less safe.

The numbers are staggering. A study conducted by Johns Hopkins found that the average ICU patient generates between 150 and 400 monitor alarms per day. Across an entire unit, that can amount to thousands of alarms per shift. Research has consistently shown that between eighty-five and ninety-nine percent of these alarms are false positives or clinically non-actionable—triggered by patient movement, sensor displacement, loose leads, or threshold settings so conservative that they flag normal physiological variation as pathology. The result is an environment of perpetual electronic noise in which the human brain, following its own well-documented survival mechanisms, learns to treat all alarms as background—including the rare ones that signal genuine emergencies.

The Joint Commission identified alarm fatigue as a national patient safety concern in 2013, and it has remained on the list every year since. Despite this designation, progress has been frustratingly slow. The reasons are partly technical—alarm systems are embedded in devices manufactured by dozens of different companies, with no universal standard for threshold calibration or alert integration—and partly cultural. There is a deeply ingrained institutional fear that reducing the number of alarms will increase the number of missed events, even though the evidence overwhelmingly demonstrates the opposite: fewer, more meaningful alarms produce better clinical responses than an avalanche of meaningless ones.

The fix has two components. The first is **customized alarm thresholds**. Instead of factory-default settings that apply the same parameters to every patient, regardless of clinical context, alarm thresholds must be calibrated to each patient's baseline. A post-operative patient on opioids has a different "normal" respiratory rate than a patient recovering from a hip replacement. A patient with chronic obstructive pulmonary disease may have a resting oxygen saturation of ninety-one percent—a level that would trigger an alarm under default settings but is perfectly normal for that individual. Calibrating alarms to the patient, rather than to the machine's factory settings, can reduce false alarm rates by as much as forty to sixty percent, according to studies conducted at Boston Medical Center and the University of California, San Francisco.

The second component is **smart escalation technology**—middleware that sits between the bedside monitor and the clinical response system, filtering, prioritizing, and routing alarms based on their clinical urgency. Under a smart escalation system, a momentary sensor displacement generates a low-priority notification logged at the central station. A sustained oxygen desaturation triggers a high-priority alert sent directly to the assigned nurse's smartphone, with automatic escalation to the charge nurse if the alert is not acknowledged within 90 seconds and to the rapid response team if not acknowledged within 3 minutes. The alarm is no longer a binary event—on or off, heard or ignored. It is a graduated signal, calibrated to the severity of the situation, with built-in redundancies that ensure someone responds. Had this system been in place on the night David Morales died, the sustained drop in his oxygen saturation would have triggered a direct alert to Karen Liu's phone at 12:22 AM. If she had not responded within ninety seconds, the charge nurse would have been notified. If neither had responded within three minutes, a rapid response team would have been dispatched. David's death required ninety minutes of unanswered alarms. A smart escalation system would have closed that window in three minutes.

3. The "Stop the Line" Authority

In the 1960s, the Toyota Motor Corporation revolutionized manufacturing with a concept so radical that Western competitors initially dismissed it as impractical: any worker on the assembly line, regardless of rank or seniority, had the authority to halt production if they identified a quality or safety concern. The concept was called "Andon," named after the cord that workers pulled to stop the line. Pulling the Andon cord was not punished. It was celebrated. Toyota's leadership understood that the cost of stopping the line for five minutes to investigate a potential defect was infinitesimal compared to the cost of allowing a defective product to reach the customer.

Healthcare operates under the opposite incentive structure. The hospital's "assembly line"—the flow of patients through admission, treatment, and discharge—is optimized for speed and throughput. Stopping the line—pausing a procedure, delaying a discharge, requesting an additional test—is implicitly discouraged by the operational culture of most institutions. It disrupts scheduling. It increases costs. It creates friction with physicians who are accustomed to having their orders executed without question. And it carries a social penalty: the nurse or technician who "stops the line" risks being labeled as overly cautious, disruptive, or insubordinate, particularly if the concern turns out to be a false alarm.

This cultural dynamic is lethal. In David's case, a respiratory therapist who happened to pass his room at 1:30 AM later told investigators that she noticed his breathing seemed unusually slow, but she did not raise the concern because he was not her assigned patient and his monitors were not showing an active alarm at that precise moment. She did not feel she had the authority—or the cultural permission—to intervene. In Elena's case, a laboratory technician who processed the critical Troponin result at 1:15 AM later said she considered calling the ER directly, as was her usual practice for critical values, but did not because the hospital's recently updated protocol routed all critical value notifications through the EHR system rather than by phone. She followed the protocol. The protocol failed.

Implementing the Stop the Line authority in healthcare requires more than a policy memo. It requires a wholesale cultural transformation in which every member of the care team—from the chief of surgery to the environmental services worker who empties the trash—understands that they have not merely the right but the obligation to raise a safety concern, and that doing so will be met with gratitude rather than retaliation. Administrators must publicly and repeatedly guarantee that stopping the line will result in a commendation, not a reprimand, even if the concern proves unfounded. The cost of a false alarm investigation is five minutes and a brief conversation. The cost of a genuine alarm is a human life.

Hospitals that have adopted formal Stop the Line programs have reported significant improvements in near-miss reporting and error detection. Virginia Mason Medical Center in Seattle, one of the earliest adopters in the United States, modeled its Patient Safety Alert System directly on Toyota's Andon cord. Staff members are trained to report any condition that could potentially harm a patient, and leadership commits to investigating every alert within a defined timeframe. The result has been a measurable reduction in serious safety events and a cultural shift in which safety concerns are treated as valuable intelligence rather than operational inconveniences.

The underlying principle at work is what Harvard organizational psychologist Amy Edmondson calls "psychological safety"—the shared belief among team members that it is safe to take interpersonal risks, to ask questions, to admit mistakes, and to raise concerns without fear of humiliation or retribution. Edmondson's research has consistently shown that teams with high psychological safety do not make fewer errors than teams without it. They report more errors—because they are willing to surface problems that would otherwise remain hidden. The distinction is crucial. The goal is not to eliminate human fallibility. That is impossible. The goal is to create an environment in which fallibility is visible, addressable, and correctable before it reaches the patient. Stop the Line is the operational expression of psychological safety: a formal mechanism that transforms a cultural principle into a concrete, actionable protocol.

4. Closed-Loop Communication Protocols

Elena Vasquez died because a critical lab result was generated, transmitted, and delivered—but never received by the person who needed it. The communication was open-loop: information was sent into the system with no mechanism to verify that it arrived at its destination and was understood by the recipient. Closed-loop communication eliminates this vulnerability by requiring confirmation at every stage of the information chain.

The concept is borrowed directly from military and aviation communication protocols, where the stakes of miscommunication are comparably lethal. When an air traffic controller issues an instruction—"Turn right heading two-seven-zero, descend and maintain flight level three-three-zero"—the pilot is required to read the instruction back verbatim, and the controller is required to confirm that the readback is correct. If the readback contains an error—"Right heading two-seven-zero, descend to three-three-zero" (omitting the word "flight level," which changes the altitude reference)—the controller catches it immediately and issues a correction. At no point is information assumed to have been received. It is verified.

In healthcare, closed-loop communication operates on two levels. The first is a **verbal closed-loop**, governed by what is known as the Read-Back Rule. Whenever a physician issues a verbal order—"Give 5 milligrams of Morphine IV"—the receiving nurse is required to repeat the order aloud: "Confirming 5 milligrams of Morphine IV." The physician then confirms: "Correct." This three-step process—order, readback, confirmation—takes approximately five seconds and eliminates the single most common source of medication errors in verbal communication: mishearing. "Fifteen" becomes distinguishable from "fifty." "Milligrams" becomes distinguishable from "micrograms." The cost of the Read-Back Rule is five seconds per order. The cost of not implementing it is measured in overdoses, underdoses, wrong-drug administrations, and deaths.

The second level is an **electronic closed-loop**, applied to the flow of critical information through the EHR system. Under an electronic closed-loop protocol, every critical lab value, imaging result, or pathology report requires a digital acknowledgment from the responsible clinician before the system permits any change in the patient's status. A discharge order cannot be finalized if an unacknowledged critical value exists in the patient's file. A transfer order cannot be processed if an outstanding diagnostic result has not been reviewed and signed off on. The system does not rely on the clinician to remember to check. It prevents the clinician from proceeding until the check is completed. This is the intervention that would have saved Elena Vasquez. Her discharge order would have been blocked by the unacknowledged Troponin result. Dr. Kim would have been forced to review the result before the system permitted Elena to leave. The heart attack would have been detected. Treatment would have begun. Elena would, in all probability, have gone home to her garden and her marigolds and her nightly phone call with Carmen.

The SBAR technique—Situation, Background, Assessment, Recommendation—provides the structural backbone for closed-loop communication during shift handoffs. By requiring the outgoing provider to organize patient information into four standardized categories, SBAR ensures that critical details are not lost in the compression of a shift change summary. Combined with the Read-Back Rule for verbal orders and electronic closed-loop for critical results, SBAR creates a communication ecosystem in which information is not merely transmitted but verified—at every handoff, at every order, at every critical juncture in the patient's care.

The Administrative Policy Roadmap

The four pillars described above are clinical and operational strategies. Their implementation, however, requires administrative infrastructure—policies, budgets, timelines, and accountability mechanisms that translate strategic intent into institutional action. The following roadmap summarizes the key policy areas that hospital leadership must address to operationalize the mitigation framework presented in this chapter.

Staffing	Acuity-Based Staffing Models	Nurse-to-patient ratios determined by clinical complexity, not fixed headcounts; mandatory fatigue management protocols
Technology	AI Sepsis Bundles & Smart Alarm Systems	Predictive algorithms to flag early infection signs 12 hours before clinical presentation; patient-specific alarm calibration; vitals-linking across devices
Communication	SBAR Handoff Protocols & Read-Back Mandates	Standardized shift handoffs with mandatory verbal and electronic closed-loop verification for all critical orders and results
Transparency	CANDOR Process	Communication and Optimal Resolution; immediate disclosure, apology, and fair settlement when errors occur, replacing adversarial litigation culture

Culture	Stop the Line Authority & Non-Punitive Reporting	Empower all staff to halt procedures for safety concerns; reward near-miss reporting; eliminate punitive responses to good-faith safety alerts

This roadmap is not exhaustive, but it addresses the five domains in which systemic reform is most urgently needed. Each policy area corresponds directly to a failure identified in David and Elena's case studies. Acuity-based staffing addresses the nurse fatigue that left Karen Liu cognitively impaired during her second consecutive overnight shift. Smart alarm technology addresses the electronic noise that drowned out David's genuine emergency. SBAR and closed-loop protocols address handoff failures that led to the loss of Elena's Troponin result. The CANDOR process addresses the institutional opacity that left Carmen Vasquez fighting for six months to learn how her mother died. And Stop the Line authority addresses the cultural silence that prevented a respiratory therapist from speaking up when she noticed David's labored breathing.

The CANDOR process—Communication and Optimal Resolution—deserves particular attention, because it addresses a dimension of hospital negligence that the other pillars do not: what happens after the error occurs. The traditional institutional response to a serious adverse event is defensive silence. The legal department advises against disclosure. The risk management team controls the flow of information. The family is told as little as possible, as carefully as possible, with the explicit goal of minimizing litigation exposure. This approach is not only ethically indefensible; it is, as the University of Michigan data demonstrate, financially counterproductive. Families who are met with transparency, empathy, and honest disclosure are significantly less likely to file lawsuits than families who are met with opacity and evasion. CANDOR replaces the adversarial model with a process of immediate disclosure, a genuine apology, a root-cause investigation conducted with family participation, and a fair financial resolution. It treats the harmed family not as a legal adversary, but as a partner in the shared project of understanding what went wrong and preventing it from happening again. Carmen Vasquez should never have needed a lawyer to learn how her mother died. Under CANDOR, she would not have.

Taken together, the five policy domains in this roadmap represent a comprehensive approach to institutional reform. No single domain is sufficient on its own. Staffing without technology leaves human vigilance unsupported. Technology without communication leaves critical information stranded in digital silos. Communication without culture leaves frontline workers afraid to speak. And culture without transparency leaves families in the dark and institutions unaccountable. The domains are interdependent. They must be implemented in concert, with sustained leadership commitment, or not at all.

From Framework to Practice

The strategies presented in this chapter are not theoretical. They are proven, evidence-based, and already operational in institutions that have committed to implementing them. The Swiss Cheese Model underpins the surgical safety checklists used in hospitals across six continents. Smart alarm systems are deployed in dozens of leading medical centers. SBAR is standard practice in the military medical system and in hospitals throughout the United Kingdom and Scandinavia. Stop the Line authority has been successfully adapted from manufacturing to healthcare in institutions from Seattle to Singapore.

The barrier to adoption is not knowledge. It is well. Every strategy in this chapter requires an upfront investment of resources, time, and political capital. Retrofitting alarm systems costs money. Acuity-based staffing costs more than fixed-ratio staffing. Training every employee in SBAR and Stop the Line authority takes hours away from clinical operations. Hospital administrators, operating under relentless pressure to reduce costs and increase throughput, face genuine tensions between financial sustainability and patient safety.

But those tensions are resolvable, and the institutions that have resolved them provide compelling proof. The University of Michigan Health System's adoption of a full disclosure and early resolution program—a forerunner of the CANDOR process—resulted in a fifty percent reduction in malpractice claims, a forty percent reduction in per-claim costs, and a sixty-one percent reduction in total liability costs over a ten-year period. Virginia Mason's implementation of its Toyota-inspired Patient Safety Alert System was accompanied by a seventy-four percent reduction in professional liability insurance costs. These are not marginal improvements. They are transformative returns on investment that dwarf the upfront cost of implementation. The financial case for patient safety is not speculative. It is proven.

What is required above all else is leadership accountability. Safety culture does not emerge from the bottom up. It is established from the top down. When hospital CEOs, chief medical officers, and boards of directors make patient safety a non-negotiable institutional priority—when they tie executive compensation to safety metrics, when they review sentinel events personally rather than delegating them to committees, when they walk the floors and ask nurses what they need—the culture shifts. When safety is treated as a line item to be managed rather than a principle to be upheld, the culture calcifies around precisely the failures this chapter has described. The framework presented here is a blueprint. But a blueprint without a builder is just paper. The builder, in every institution, is the leader.

In the next chapter, we turn to the most overlooked and most powerful layer of defense in the entire safety architecture: the patient and the patient's family. Because the most sensitive alarm in any hospital room is not electronic. It is the person sitting at the bedside who knows the patient better than any monitor ever will—and who, with the right tools and the right support, can become the final, irreplaceable safeguard against the failures that no system, however well designed, can entirely eliminate.

CHAPTER FIVE

The Advocacy Layer

Engaging the Family as the Final Line of Defense

Dr. Dorathy U. Ubiam

"Nobody knows a patient the way their family does. The chart tells you the numbers. The family tells you the person."

— Sorrel King, Patient Safety Advocate

In every chapter of this book so far, we have discussed safety in institutional terms—systems, protocols, technologies, policies. We have examined how hospitals can redesign their alarm architectures, restructure their communication pathways, and cultivate cultures of psychological safety. These are essential interventions. Without them, the silent epidemic of hospital negligence will continue unchecked. But there is one safety mechanism that no chapter on institutional reform can fully address, because it does not live inside the institution. It lives at the bedside. It arrives in the morning with a bag of clean clothes and a container of homemade soup. It stays past visiting hours because something does not feel right. It is the patient's family.

The family—the spouse, the parent, the adult child, the friend who might as well be family—is the most underutilized, most overlooked, and potentially most powerful safety resource in the entire healthcare system. They are the ones who know that the patient normally sleeps on his left side and has never snored like that before. They are the ones who notice that her color has changed since this morning, that she is answering questions more slowly than usual, that something in his eyes looks different in a way that no vital sign can quantify. They carry a lifetime of baseline data that no electronic health record can replicate, and they are present at the bedside during the hours when staffing is thinnest and attention is most stretched—the overnight shifts, the holiday weekends, the shift changes when information falls between the cracks.

And yet, in the traditional culture of hospital medicine, the family is treated not as a partner in care, but as a visitor—tolerated, managed, and, when they become insistent, labeled. The label most families fear is the one that silences them: "difficult."

The "Difficult" Family

There is a word that appears in nursing notes and physician progress reports with a frequency that should alarm anyone concerned with patient safety. The word is "difficult." It is applied, almost always, to family members who ask too many questions, who challenge clinical decisions, who press the call button more often than the staff considers appropriate, or who insist that something is wrong when the monitors say everything is fine. In the shorthand of hospital culture, a "difficult" family is one that refuses to defer quietly to professional authority.

The irony is devastating. The behaviors that earn the label "difficult"—vigilance, persistence, refusal to be reassured by vague answers—are precisely the behaviors that patient safety advocates identify as the most effective defenses against medical error. A family member who insists that the patient's breathing sounds different is performing, in real time, the kind of clinical surveillance that alarm systems and monitoring protocols are designed to replicate. A family member who asks why the lab results have not come back is doing exactly what Elena Vasquez's care team failed to do on the night she was discharged. A family member who refuses to leave the bedside because something feels wrong is exercising a form of situational awareness that no staffing ratio can guarantee.

The problem is not that families are difficult. The problem is that the healthcare system has not been designed to accommodate their participation. Hospital workflows are built around the assumption that clinical decisions are made by credentialed professionals and executed by trained staff, with the patient and family as passive recipients. When a family member steps outside that role—when they question, challenge, or advocate—they disrupt the workflow. And in an environment where workflow disruptions can cascade into delays, missed tasks, and interpersonal friction, the institutional reflex is to contain the disruption rather than listen to it.

This reflex has lethal consequences. Research published in the Joint Commission Journal on Quality and Patient Safety found that in a significant percentage of rapid response events—situations in which a patient's condition deteriorated rapidly enough to require emergency intervention—family members had expressed concerns about the patient's condition hours before the clinical team recognized the problem. In many of those cases, the family's concerns were dismissed, minimized, or attributed to anxiety. The patient safety literature is clear: when families speak up and are heard, outcomes improve. When families speak up and are silenced, patients die.

Condition H: Help Is a Phone Call Away

In 2001, a twenty-month-old girl named Josie King was admitted to Johns Hopkins Hospital for treatment of first- and second-degree burns sustained in a bath. Her injuries, while serious, were not life-threatening. She was expected to make a full recovery. Over the course of her hospitalization, a series of medical errors—including a failure to address her increasing thirst and lethargy, which were signs of severe dehydration caused by a medication interaction—led to cardiac arrest. Josie died on February 22, 2001. She was eighteen months old.

Josie's mother, Sorrel King, would later describe the hours before her daughter's death as a nightmare of escalating alarm and institutional indifference. She told nurses repeatedly that something was wrong—that Josie was lethargic, that her eyes looked sunken, that she was desperately thirsty. Her concerns were noted. They were not acted upon. When she confronted the medical team directly, she was reassured that everything was under control. By the time the clinical staff recognized the severity of Josie's condition, it was too late.

Sorrel King's response to her daughter's death was extraordinary. Rather than retreating into private grief, she partnered with Johns Hopkins to create one of the most important patient safety innovations of the past quarter century: Condition H. Using a portion of the settlement from Josie's case, King established the Josie King Foundation, which has since funded patient safety programs at hospitals across the country and become one of the most influential voices in the family advocacy movement. King's central insight was simple and revolutionary: the mother who had tried to save her daughter's life by speaking up should not have been the only person in the room who lacked the formal authority to be heard.

Since its introduction at Johns Hopkins, Condition H has been adopted by hundreds of hospitals in the United States and internationally, under various names—Condition H, Condition Help, Patient and Family Activated Rapid Response, or simply Family Call. The specific nomenclature varies, but the principle is universal: the family is granted formal, institutional authority to escalate a clinical concern when they believe the standard channels have failed.

Condition H—the H stands for "Help"—is a policy that allows patients and their family members to call a Rapid Response Team directly, bypassing the usual chain of clinical command. At hospitals that have implemented Condition H, a dedicated phone number is posted in every patient room, printed on every admission packet, and explained to every family at intake. If a patient or family member believes that the medical staff is not responding adequately to a deteriorating condition—if they feel their concerns are being dismissed, if they observe a change that the clinical team has not acknowledged, if something simply does not feel right— they can pick up the phone and activate a Rapid Response Team themselves.

The response is immediate. A team of critical care specialists arrives at the bedside, independently assesses the patient, and intervenes as necessary. The family's activation of Condition H is treated with the same clinical seriousness as an alarm triggered by a bedside monitor. There is no intermediary. There is no filter. There is no nurse or resident who must first agree that the concern is legitimate before escalation occurs. The family's judgment is, by policy, sufficient.

The results have been striking. Hospitals that have adopted Condition H report that family activations, while infrequent— typically a handful per month—have in multiple documented cases, identified genuine clinical deterioration that the care team had not yet recognized. In some instances, Condition H activations have led to transfers to the intensive care unit, emergency surgical interventions, and medication adjustments that prevented imminent harm. False activations—cases in which the family's concern did not correspond to a clinical problem—are treated not as inconveniences but as opportunities to improve communication between the family and the care team. The cost of a false activation is a five-minute assessment. The cost of a missed genuine activation is, once again, a human life.

Consider how Condition H might have altered the trajectory of David Morales's case. Maria Morales was not at the hospital on the night David died—he was recovering from routine surgery, and there was no reason for her to stay overnight. But in cases where family members are present, Condition H provides a failsafe that operates independently of the alarm system, the staffing ratio, and the nursing assignment. If Maria had been there, if she had noticed the change in David's breathing, if she had pressed the call button and been told that the monitors showed no cause for concern, she could have picked up the phone and activated Condition H. A fresh pair of clinical eyes would have arrived at the bedside. David's respiratory depression would have been identified. The PCA pump would have been paused. He would, in all likelihood, be alive.

Empowering Families to Speak Up

Condition H is a structural intervention—a policy that creates a formal pathway for family advocacy. But policies are only as effective as the culture that surrounds them. A Condition H phone number posted on a wall is meaningless if the family is too intimidated, too deferential, or too afraid of being labeled "difficult" to pick up the phone. Empowering families to speak up requires more than a hotline. It requires a fundamental shift in how hospitals communicate with the people who love the patients they treat.

That shift begins at admission. The first hours of a hospitalization set the tone for the entire relationship between the care team and the family. If the family is treated from the outset as a nuisance to be managed—if their questions are met with impatience, if they are shooed out of the room during rounds, if they are given vague assurances instead of honest information—they will learn, quickly, that their role is to be quiet and grateful. If, on the other hand, the family is welcomed as a partner in care—if their knowledge of the patient is actively solicited, if they are included in bedside rounds, if they are told explicitly that their observations matter and that speaking up is not just permitted but encouraged—they will become the most attentive, most motivated, and most continuously present monitors in the hospital.

Several hospitals have formalized this approach through Patient and Family Advisory Councils—standing committees composed of former patients, family members, and community representatives who provide input on hospital policies, quality improvement initiatives, and patient experience programs. These councils are not symbolic. At institutions that take them seriously, they have driven meaningful changes in visiting hours policies, discharge planning processes, and communication protocols. They bring a perspective that no clinician, however empathetic, can fully replicate: the perspective of someone who has lain in the bed, or sat beside it, and experienced the healthcare system not as a professional but as a human being whose life, or whose loved one's life, depends on it.

One of the most consequential changes driven by family advocacy has been the expansion of visiting hours. For decades, hospitals operated under rigid visiting schedules—typically a few hours in the afternoon and early evening—designed around institutional convenience rather than patient welfare. The rationale was that unrestricted family presence would disrupt clinical workflows, interfere with rest, and create infection control challenges. The evidence tells a different story. Research published in Critical Care Medicine found that flexible visiting policies in intensive care units were associated with reduced patient anxiety, lower incidence of delirium, improved family satisfaction, and no measurable increase in infection rates. More importantly, unrestricted family presence means more hours of human observation at the bedside—more opportunities for the kind of vigilance that saved patients in Condition H activations and that was absent on the night David Morales died. The family is not a disruption to care. The family is care.

Another critical development is the Open Notes movement, which gives patients and families real-time electronic access to their medical records, including physician progress notes, lab results, and imaging reports. Launched as a pilot at three major medical centers in 2010 and now mandated under federal information-blocking rules, Open Notes fundamentally rebalances the information asymmetry between the care team and the family. When Carmen Vasquez needed an attorney and a subpoena to see her mother's Troponin result, she was operating under the old paradigm—one in which the medical record belonged to the institution. Under Open Notes, Elena's critical lab value would have been visible to her family in real time, on their phones, the moment it was finalized. Whether anyone at the hospital was checking the result, the family could have been as well. The information would not have vanished into an unmonitored inbox. It would have been visible to the people with the strongest incentive in the world to act on it: those who loved her.

But the most critical form of family empowerment is not institutional. It is informational. Families cannot advocate effectively if they do not know what to look for, what questions to ask, or what responses to expect. The remainder of this chapter is addressed directly to you—the reader who may one day find yourself at a loved one's bedside, wondering whether what you are seeing is normal, wondering whether you should say something, wondering whether your concern will be taken seriously or brushed aside.

What You Can Do: A Guide for the Bedside Advocate

Know your patient's baseline. Before surgery, before admission, before the first dose of medication, take a mental inventory of what "normal" looks like for your loved one. How do they normally breathe when they sleep? What color is their skin at rest? How quickly do they respond to questions? How alert are their eyes? You know these things intuitively because you have observed this person for years. The clinical team has observed them for hours. Your baseline data is irreplaceable, and you should communicate it clearly at admission. Tell the nurse: "He normally snores lightly but never breathes like that." Tell the doctor: "She's usually very sharp and responsive, so the fact that she seems confused concerns me." This information can be the early warning signal that no monitor can provide.

Ask about the plan. You have the right to understand what is being done, why it is being done, and what the expected outcome is. You do not need a medical degree to ask: "What tests have been ordered, and when should we expect the results?" "What medications is he receiving, and what are the possible side effects?" "What are the signs I should watch for that would indicate a problem?" These questions are not intrusive. They are the questions that informed, engaged care partners ask, and any healthcare professional who bristles at them is revealing more about the institution's culture than about your behavior.

Trust your instincts. If something does not feel right, say so. Do not wait for the monitors to confirm what your eyes and your gut are telling you. The clinical term for what you are experiencing is "clinical intuition by proxy"—the ability of a close observer to detect subtle changes that precede measurable decline. You do not need to diagnose the problem. You need only to name what you are seeing: "Her breathing has changed." "He doesn't look right to me." "Something is different from this morning." If your concern is dismissed, escalate. Ask for the charge nurse. Ask for the attending physician. And if you are still not satisfied, ask about Condition H or the hospital's equivalent rapid response activation policy.

Document everything. Keep a simple log—a notebook, a phone note, a voice memo—of key events, times, and observations during your loved one's hospitalization. Note when medications are given, when vital signs are checked, when you raise a concern, and how it is responded to. This is not adversarial. It is prudent. If something goes wrong, your contemporaneous notes will be the most reliable record of what happened and when. If nothing goes wrong, the log costs you nothing. But in the event of an adverse outcome, the difference between a family that can reconstruct a timeline and one that relies on shock-blurred memory can be the difference between accountability and silence.

Be present during transitions. Shift changes and handoffs are the most dangerous moments in a hospitalization, as we have seen in Elena's case. If you can be present during a shift change—typically around seven in the morning and seven in the evening—you can serve as a living bridge of continuity between the outgoing and incoming teams. You can ensure that the concerns you raised with the day nurse are communicated to the night nurse. You can confirm that pending test results have been flagged. You can provide the incoming team with the baseline information that the outgoing team may not have had time to fully convey. Your presence during these transitions does not guarantee that nothing will fall through the cracks. But it significantly reduces crack size.

Know when to escalate. If you have raised a concern with the bedside nurse and been reassured, but your concern persists, ask to speak with the charge nurse. If the charge nurse's response does not satisfy you, ask for the attending physician. If the attending physician is unavailable or unresponsive, ask the nursing station about the hospital's Rapid Response Team activation process. In hospitals with Condition H, call the posted number. At no point in this escalation chain should you feel that you are being unreasonable. You are exercising a right endorsed by every major patient safety organization worldwide. The worst that can happen if you are wrong is a brief clinical assessment that confirms your loved one is stable. The worst that can happen if you are right and remain silent is something no family should ever have to live with.

The Fear of Retaliation

It would be dishonest to write a chapter on family advocacy without acknowledging the fear that prevents many families from speaking up: the fear that challenging the clinical team will result in worse care for their loved one. This fear is not irrational. It is rooted in the power asymmetry that defines the hospital experience. The patient is vulnerable, dependent, and physically confined. The care team holds all the expertise, access, and authority. A family that antagonizes the staff risks, or believes it risks, subtle shifts in attentiveness—a slightly longer wait for pain medication, a slightly less thorough check during rounds, an unspoken downgrade in the quality of care that can never be proven but is always felt.

This fear is real, and the healthcare system has an obligation to eliminate it. That obligation falls primarily on institutional leadership rather than on the family. Hospitals must train staff explicitly in the principle that family advocacy is a safety asset, not a behavioral problem. Nursing education must include modules on responding constructively to family concerns, with role-playing exercises that teach de-escalation, active listening, and the clinical integration of family observations. Performance evaluations for nursing and medical staff must include metrics on family communication and responsiveness. And hospital administrations must publicly and unambiguously establish a zero-tolerance policy against retaliatory behavior toward patients or families who raise safety concerns.

These measures will not eliminate the power asymmetry overnight. But they will begin to shift the culture from one in which the family's role is to watch quietly and hope for the best, to one in which the family's role is to participate actively in the shared project of keeping the patient alive.

In the meantime, families navigating this reality should know several things. First, the vast majority of nurses and physicians welcome engaged, informed family members. Surveys of nursing staff consistently show that while some clinicians find frequent family questioning stressful, the majority recognize it as a sign of investment in the patient's care, and many report that family observations have alerted them to clinical changes they might otherwise have missed. Second, framing your advocacy in collaborative rather than adversarial language reduces friction without reducing effectiveness. "I'm noticing something that concerns me, and I want to make sure we're on the same page" communicates the same urgency as "Why isn't anyone doing anything?" while preserving the working relationship that your loved one depends on. Third, and most importantly: if you are genuinely concerned about your loved one's safety, the risk of speaking up is always—always—smaller than the risk of staying silent. There has never been a family that wished they had said less at the bedside of someone they lost.

The Last Line of Defense

Every safety system described in this book—the Swiss Cheese Model, smart alarms, closed-loop communication, Stop the Line authority—is designed by humans, implemented by humans, and operated by humans. Every one of them, therefore, has the potential to fail. The alarm can malfunction. The protocol can be bypassed. The software can glitch. The culture can lapse. When all of these layers have holes, and the accident trajectory begins to align, there is one final layer that no institution can control and no technology can replace: the person at the bedside who knows the patient, who loves the patient, and who will not leave until they are certain the patient is safe.

That person is you. And this book's deepest conviction is that you are not a visitor. You are not a bystander. You are not an obstacle to clinical efficiency. You are the last line of defense. And when you speak up—when you refuse to be reassured by a monitor reading that contradicts what your eyes are telling you, when you pick up the phone and call a Rapid Response Team because something does not feel right, when you trust the knowledge you carry about the person in that bed over the algorithms and protocols that do not know them the way you do—you are performing the most important act of advocacy in the entire healthcare system.

David Morales did not have someone at his bedside on the night he died. Elena Vasquez was sent home before her family could process what was happening. Neither of them had an advocate who was empowered to intervene. This chapter exists so that the next patient—the one who will be admitted tomorrow, to a hospital not far from where you are reading this—will.

In the next chapter, we look forward. We examine the emerging technologies and systemic reforms—from artificial intelligence to the Black Box approach borrowed from aviation—that represent the future of patient safety. The path to zero harm is long, but it is no longer uncharted. The tools exist. The evidence exists. The question that remains is whether we have the collective courage to use them.

CHAPTER SIX

The Path to Zero Harm

Dr. Dorathy U. Ubiam

*"In God we trust. All others must
bring data."*

— W. Edwards Deming

The preceding chapters have mapped the problem and outlined the available solutions—proven, evidence-based strategies that could dramatically reduce preventable deaths in hospitals if implemented with institutional commitment. But this book would be incomplete if it stopped at the present. The landscape of patient safety is not static. It is being reshaped, in real time, by two forces that have the potential to fundamentally alter the relationship between human vulnerability and medical harm: artificial intelligence and the systematic study of failure.

This chapter looks forward. It examines two emerging approaches that represent the frontier of patient safety innovation. The first—the "Black Box" approach—borrows directly from the aviation industry's most iconic safety tool to create a new paradigm for understanding how and why things go wrong in the operating room and beyond. The second—AI-driven real-time surveillance—deploys machine learning algorithms to detect the earliest signals of patient deterioration, often hours before a human clinician could identify them. Together, they represent a vision of healthcare in which preventable death is not merely reduced, but systematically driven toward zero.

That vision is not utopian. It is engineering. And the engineers are already at work.

The "Black Box" Approach: Learning From Every Failure

On the morning of June 1, 2009, Air France Flight 447 departed Rio de Janeiro for Paris with 228 people on board. Three hours and thirty minutes into the flight, the aircraft encountered a high-altitude storm over the Atlantic Ocean. Ice crystals blocked the plane's external airspeed sensors, causing the autopilot to disconnect and presenting the pilots with a cascade of contradictory instrument readings. Over the next four minutes and twenty-four seconds, the flight crew made a series of decisions that, in hindsight, were catastrophic. The co-pilot, disoriented by the conflicting data, pulled back on the control stick, sending the aircraft into an aerodynamic stall from which it never recovered. Flight 447 plunged into the Atlantic, killing everyone on board.

It took nearly two years to locate the wreckage on the ocean floor, at a depth of nearly thirteen thousand feet. But when investigators recovered the aircraft's flight data recorder and cockpit voice recorder—the devices commonly known as black boxes—they were able to reconstruct, second by second, every input, every instrument reading, every word spoken in the cockpit during those final four minutes. The data revealed not pilot incompetence, but a systemic failure: inadequate training for manual flight at high altitude, confusing cockpit automation logic that provided contradictory stall warnings, and a communication breakdown between the two co-pilots and the captain that prevented the crew from recognizing and correcting the stall in time.

The investigation that followed did not result in criminal charges against the crew. It resulted in sweeping changes to pilot training protocols, cockpit interface design, and airspeed sensor technology across the global aviation fleet. The black box made this possible. Without it, the cause of Flight 447's loss would have remained a mystery, buried beneath twelve thousand feet of water. With it, the aviation industry gained the precise, granular, second-by-second understanding of failure that is the prerequisite for systemic improvement.

Now ask the obvious question: Why doesn't healthcare have a black box?

The Operating Room Black Box

In 2014, a surgical safety researcher named Dr. Teodor Grantcharov at the University of Toronto asked precisely that question and began building the answer. Grantcharov, himself a practicing surgeon, had spent years studying surgical error and had arrived at a frustrating conclusion: the healthcare system's capacity to learn from its mistakes was fundamentally limited by the quality of the data it collected about those mistakes. Incident reports were filed days or weeks after the event, relying on the memories of participants who were emotionally distressed and cognitively biased. Root cause analyses were conducted by committee, based on chart reviews and interviews that inevitably reflected the perspectives and self-interest of the people involved. The operating room—the site of some of medicine's most consequential and most preventable failures—was, from a data perspective, a black hole.

Grantcharov's solution was to bring the aviation industry's black box concept into the surgical suite. He developed a system that captures, synchronizes, and records multiple data streams during every surgical procedure: high-definition video of the operative field, audio from the room including all team communications, real-time physiological data from the patient's monitors (heart rate, blood pressure, oxygen saturation, end-tidal carbon dioxide), and data from surgical instruments and devices. The result is a comprehensive, objective, time-stamped record of everything that happens during an operation—a record that can be reviewed, analyzed, and learned from with the same forensic precision that aviation investigators bring to a crash.

The implications are profound. Consider a surgical complication—an unexpected hemorrhage, a bowel perforation, an anesthesia-related cardiac event. Under the traditional model, the surgeon files an incident report based on recollection, the anesthesiologist files a separate report, and the nursing staff files a third. These reports are inevitably incomplete, sometimes contradictory, and always filtered through the lens of hindsight and self-preservation. The root cause analysis that follows is, at best, an educated reconstruction. At worst, it is a negotiated narrative in which institutional politics determine which facts are emphasized and which are quietly omitted.

Under the black box model, the review begins not with recollection but with data. The team can watch exactly what happened, listen to exactly what was said, and correlate team behavior with patient physiology in real time. Did the hemorrhage begin before or after a particular instrument was introduced? Did the anesthesiologist alert the surgeon to the dropping blood pressure, and if so, how quickly did the surgeon respond? Was the surgical safety checklist actually performed, or was it skipped? Was a critical piece of information communicated clearly, or was it lost in the ambient noise of the operating room? The black box answers these questions with evidence, not testimony.

Crucially, the black box system developed by Grantcharov is designed to be non-punitive. Recordings are anonymized and governed by strict confidentiality protocols. They are used for quality improvement and education, not for disciplinary action or litigation. This distinction is essential. If surgical teams believed that the black box was a surveillance tool—a mechanism for institutional oversight or legal discovery—they would resist it, and rightly so. The entire value of the system depends on the same cultural condition that underpins every other safety innovation described in this book: trust. The black box must be perceived as a learning tool, not a weapon. Aviation achieved this decades ago through legislation that protects flight recorder data from use in criminal prosecution except under narrowly defined circumstances. Healthcare must follow the same path, establishing legal protections for black box data that encourage transparency rather than defensiveness.

Early results from hospitals piloting the surgical black box have been encouraging. Analysis of recorded procedures has identified patterns of error that were invisible under the traditional reporting model: communication breakdowns during critical moments that were not captured in incident reports, equipment malfunctions attributed to user error, and near-misses that were never reported because the team was unaware they had occurred. In one published study, the black box identified a relevant safety event in nearly one-third of all recorded procedures—a rate dramatically higher than that suggested by traditional incident reporting. The operating room, it turns out, is not as safe as the incident reports claim. The black box reveals the gap between what hospitals report and what actually happens, and in that gap lies the opportunity for transformative improvement.

The black box also has the potential to transform surgical education. Currently, surgical training relies heavily on the apprenticeship model: the trainee watches the attending surgeon, assists in progressively more complex procedures, and gradually assumes independent responsibility under supervision. The quality of this training depends almost entirely on the individual attending surgeon's willingness and ability to teach, to debrief, and to acknowledge their own errors openly. With a black box, every procedure becomes a teaching case. Surgical residents can review recordings of both successful and unsuccessful operations, analyzing team dynamics, communication patterns, and decision-making processes with the benefit of replay, slow motion, and the overlay of physiological data that contextualizes every action. The morbidity and mortality conference—the traditional forum for reviewing surgical complications—is transformed from a recollection-based discussion into an evidence-based analysis. The surgeon no longer has to remember what happened. The data shows it.

The cultural parallel to aviation is instructive. In the early decades of commercial flight, pilots resisted cockpit voice recorders with the same intensity that some surgeons resist the operating room black box today. They viewed recording as surveillance. They feared that their words would be taken out of context, used against them in litigation, or weaponized by management. It took decades of legislative protection, demonstrated benefit, and cultural evolution for the cockpit voice recorder to be accepted as a standard, non-negotiable component of flight safety. Today, no pilot would fly without one. The black box is not the enemy of the pilot. It is the pilot's most powerful advocate in the aftermath of an incident, providing objective data that often exonerates the crew and reveals systemic failures that would otherwise have been attributed to individual error. The surgical black box will follow the same trajectory. Resistance will give way to acceptance, and acceptance will give way to the realization that operating without a black box was never really operating safely at all.

AI and Real-Time Surveillance: Predicting Harm Before It Happens

If the black box addresses the question of how we learn from failures after they occur, artificial intelligence addresses an even more ambitious question: Can we predict failures before they happen?

The premise is straightforward. Every patient in a modern hospital generates a continuous stream of physiological data: heart rate, blood pressure, respiratory rate, oxygen saturation, temperature, and dozens of laboratory values over the course of a hospitalization. This data is currently monitored by human beings—nurses and physicians who review vital signs at scheduled intervals, scan lab results as they arrive, and respond to alarm thresholds that, as we have seen, are set at levels so conservative that they generate more noise than signal. The human monitoring model is limited by three fundamental constraints: attention, time, and pattern recognition. A nurse monitoring six patients cannot watch all six simultaneously. A physician reviewing lab results cannot hold in working memory the dozens of subtle, interacting trends that may signal a developing crisis. And no human being, however experienced, can reliably detect the faint, multi-variable patterns that distinguish the early stages of sepsis from the normal post-operative inflammatory response.

Artificial intelligence has none of these limitations. Machine learning algorithms can process the entire data stream from every patient on a unit simultaneously, continuously, and without fatigue. They can detect patterns in the interaction between variables—a subtle combination of rising heart rate, falling blood pressure, and marginally elevated white blood cell count—that would be invisible to even the most attentive clinician because the individual values remain within normal ranges. The algorithm does not wait for a threshold to be crossed. It identifies the trajectory and projects the likely destination.

The Sepsis Challenge

Nowhere is this capability more consequential than in the detection of sepsis—a life-threatening condition in which the body's immune response to an infection spirals out of control, damaging organs and, in severe cases, causing death. Sepsis kills more than 250,000 Americans annually and is the leading cause of death in hospital intensive care units. The critical variable in sepsis survival is time: every hour of delay in initiating appropriate antibiotic therapy increases mortality by approximately seven to eight percent. A patient who receives antibiotics within the first hour of sepsis onset has a dramatically better survival rate than a patient who receives them six hours later. The difference is often the difference between life and death.

The problem is detection. The early signs of sepsis are notoriously subtle and overlap with the symptoms of dozens of benign conditions. A post-operative patient with a mild fever, a slightly elevated heart rate, and a marginally abnormal white blood cell count may be developing sepsis— or may simply be experiencing the normal inflammatory response to surgery. The clinical presentation is ambiguous, and waiting for clarity can be fatal. Traditional clinical surveillance relies on human judgment applied at intervals, using screening tools with limited sensitivity and high false-positive rates. The result is a diagnostic challenge that perfectly illustrates the limitations of human monitoring in a complex, time-sensitive environment.

AI-powered sepsis detection systems approach this challenge differently. Rather than relying on a static set of screening criteria applied at fixed intervals, they continuously analyze the full spectrum of available patient data—vital signs, laboratory values, medication administration records, nursing assessments, even the timing and content of clinical notes—to generate a real-time probability score for sepsis onset. When the algorithm identifies a patient whose data pattern is consistent with early sepsis, it generates an alert to the clinical team, typically hours before the condition would become clinically apparent through traditional monitoring.

The results from early deployments have been significant. A study published in Nature Medicine reported that an AI-based early warning system deployed at a major academic medical center predicted sepsis onset up to 12 hours before clinical recognition, with sensitivity and specificity that outperformed those of traditional screening tools. Hospitals that have implemented similar systems have reported measurable reductions in sepsis-related mortality, length of stay, and cost of care. The algorithm does not replace the clinician. It augments the clinician, providing a layer of surveillance that operates at a speed, scale, and level of pattern recognition that the human brain, however brilliant, simply cannot match.

Beyond Sepsis: The Broader Vision

Sepsis is the proving ground, but it is not the destination. The same AI capabilities that enable early sepsis detection can be applied to virtually any condition in which early recognition affects outcome. Cardiac arrest prediction. Respiratory failure. Acute kidney injury. Decompensation in post-surgical patients. Opioid-induced respiratory depression—the condition that killed David Morales. In each of these scenarios, the physiological data preceding the crisis contains signals detectable by machine learning algorithms but invisible to human observers operating under the constraints of time, attention, and cognitive load.

Imagine a hospital in which every patient's data stream is continuously analyzed by an AI system that integrates vital signs, lab values, medication records, and nursing observations into a single, unified risk score. The score is displayed on a dashboard visible to the charge nurse, updated in real time, and color-coded by urgency: green for stable, yellow for watch, red for immediate intervention required. The system does not generate hundreds of false alarms per shift. It generates a handful of high-confidence, high-specificity alerts that the clinical team has learned to trust, because the algorithm has been trained on millions of patient encounters and validated against thousands of outcomes. When the dashboard turns red, someone comes. *Every time.*

This is not a fantasy. Elements of this system are already operational in hospitals across the United States, the United Kingdom, and Israel. The technology is maturing rapidly, driven by advances in machine learning, the increasing digitization of clinical data, and the growing recognition that human monitoring alone is insufficient for the complexity and acuity of modern hospital care. The question is no longer whether AI will play a central role in patient safety. The question is how quickly hospitals will adopt it, and whether the adoption will be driven by proactive leadership or by the accumulation of preventable deaths that finally become too numerous to ignore.

What These Tools Would Have Meant

It is worth pausing to apply these emerging technologies to the cases that have anchored this book. If a surgical black box had been standard in David Morales's hospital, the post-operative period would have been continuously recorded. The gradual decline in his respiratory rate, the timing and duration of each alarm, the eleven-minute gap between alarm and response, the remote acknowledgment without bedside verification—all of it would have been captured with forensic precision. The root cause analysis would not have relied on Karen Liu's traumatized recollection or the hospital's self-interested internal review. It would have been based on data. And the resulting recommendations would have carried the weight of objective evidence rather than the ambiguity of institutional narrative.

If an AI surveillance system had been monitoring David's data stream, the algorithm would have detected the pattern of declining respiratory rate and oxygen saturation long before the pulse oximeter alarm triggered at 12:22 AM. It would have identified the trajectory—not just the isolated data point—and generated a high-confidence alert to the clinical team, potentially as early as 11:00 PM, when his respiratory rate first began to slow. The PCA pump could have been paused. A clinical assessment could have been performed. David would have woken up on Wednesday morning with a sore knee and a story about hospital food, and Sofia's get-well card would have been delivered.

If the same AI system had been operating in Elena Vasquez's emergency department, it would have flagged her critical Troponin result the moment it was finalized, regardless of which physician was logged in, regardless of the shift change, regardless of the EHR notification routing architecture. The algorithm does not care about shift changes. It does not log out. It does not go home. It continuously monitors every piece of data for every patient and escalates every critical finding until a responsible clinician acknowledges it. Elena's heart attack would have been detected. She would have been treated. She would have gone home to her garden.

The Ethical Horizon

No discussion of AI in healthcare would be responsible without addressing the ethical complexities that accompany it. Machine learning algorithms are trained on historical data, which reflects historical biases. If the training data overrepresents certain demographic groups and underrepresents others, the algorithm's predictive accuracy will be correspondingly skewed—potentially generating fewer alerts for patients from underrepresented populations, precisely the populations that already experience disparate outcomes in the healthcare system. Algorithmic bias is not a theoretical concern. It has been documented in multiple clinical AI systems and must be addressed through rigorous, ongoing validation across diverse patient populations.

There is also the question of clinical autonomy. Physicians are trained to exercise independent judgment based on their education, experience, and direct observation of the patient. An AI system that generates alerts based on statistical patterns may, in some cases, conflict with the clinician's assessment. How should that conflict be resolved? The answer, grounded in the evidence to date, is that AI should augment clinical judgment, not override it. The algorithm provides information. The clinician makes the decision. But the clinician must take the information seriously, and the institution must create a culture in which dismissing an AI alert without documented clinical reasoning is treated with the same gravity as ignoring any other safety signal.

The ideal model is one of complementarity rather than competition. The AI excels at what humans cannot do well: continuous monitoring without fatigue, simultaneous pattern detection across dozens of variables, and probabilistic reasoning at a speed and scale that exceed human cognitive capacity. The human excels at what the AI cannot do at all:

interpreting context, exercising ethical judgment, communicating with empathy, and integrating the qualitative, unquantifiable dimensions of patient care—the look in a patient's eyes, the tone of a family member's voice, the clinical intuition that something is wrong before any data point confirms it. Neither the human nor the machine is sufficient on its own. Together, they form a monitoring partnership that is more capable, more attentive, and more resilient than either could be on its own. The future of patient safety is not the replacement of the clinician by the algorithm. It is the liberation of the clinician from tasks that the algorithm performs better, so that the clinician can devote more time, attention, and presence to the tasks only a human being can perform: listening, observing, touching, and caring.

Privacy, too, demands careful attention. The continuous recording of operating room activity by a surgical black box raises legitimate concerns about surveillance, consent, and potential data misuse. Patients and surgical teams alike must be informed about recording practices and must have confidence that the data will be used exclusively for quality improvement and education, never for punitive or commercial purposes. The legal and regulatory frameworks governing this data are still developing, and healthcare institutions have an obligation to engage proactively in shaping them—establishing protections that encourage transparency rather than the defensive opacity that currently characterizes so much of the industry's relationship with its own failures.

The Road Ahead

The path to zero harm is long. It is not a destination that any healthcare system will reach tomorrow, or next year, or perhaps even in the next decade. But it is a path, not a mirage. The tools described in this chapter—the surgical black box, AI-driven surveillance, predictive analytics—are not speculative technologies awaiting some future breakthrough. They exist now. They are operational now. They are producing measurable results now. What separates the hospitals that use them from those that do not is the same variable that has separated progress from stagnation throughout this entire book: the will to act.

The aviation industry did not achieve its extraordinary safety record by waiting for perfection. It achieved it through the relentless, iterative, data-driven study of failure—one crash, one near-miss, one black box recording at a time. Each failure was treated not as an embarrassment to be concealed, but as an invaluable source of intelligence to be mined, shared, and applied across the entire industry. Healthcare must make the same commitment. Every preventable death is a lesson. Every near-miss is a gift. And every tool that helps us see more clearly what went wrong, and why, and how to prevent it from happening again, brings us one step closer to the day when the silent epidemic is no longer silent and no longer an epidemic.

This book has been, from the first page, an argument for seeing what we have been trained not to see. The empty bed. The ignored alarm. The lab result vanished into an unmonitored inbox. The nurse was too exhausted to trust her own instincts. The family that was too afraid to speak up. The system that classified every one of these failures as an acceptable cost of doing business. These are not inevitable features of modern medicine. They are design choices—choices made by institutions, by administrators, by policymakers, and by a culture that has decided, whether consciously or not, that a certain number of preventable deaths is a price worth paying for efficiency, throughput, and the preservation of professional hierarchies that were never designed with patient safety as their primary concern.

This book rejects that bargain. It insists that every preventable death is one too many, that the tools to prevent them exist, and that the only acceptable trajectory is toward zero. The black box shows us what really happens. Artificial intelligence shows us what is about to happen. The Swiss Cheese Model, the alarm reforms, the closed-loop protocols, the Stop the Line authority, the family advocates at the bedside—each of these is a layer of defense, and together they form an architecture of safety that is more robust, more resilient, and more humane than anything the healthcare industry has built to date. The path to zero harm begins with a single, radical act of honesty: admitting that the current system is not good enough, and committing to build one that is.

David Morales deserved better. Elena Vasquez deserved better. Josie King deserved better. And the bed in room 417 waits for the day when the patient who occupies it is still alive in the morning, recovering, healing, and waiting for his daughter to deliver a get-well card with a lopsided soccer ball on the front.

APPENDIX A

The Patient's Bill of Rights

Every patient admitted to a hospital has fundamental rights that exist regardless of diagnosis, insurance status, or institutional policy. These rights are grounded in medical ethics, federal and state law, and the basic principle that every human being deserves safe, transparent, and dignified care. This Bill of Rights is presented as both a reference for patients and families and a standard against which institutional performance can be measured.

1. The Right to Safe Care

You have the right to receive care that meets established safety standards. This includes the right to be treated by qualified, credentialed professionals; the right to have your identity, surgical site, and procedure verified before any intervention; and the right to be monitored with appropriate technology and adequate staffing throughout your hospitalization.

2. The Right to Full Information

You have the right to receive complete, accurate, and timely information about your diagnosis, treatment options, risks, benefits, and alternatives, communicated in language you can understand. This includes the right to access your medical records in real time, including lab results, imaging reports, and physician notes, as guaranteed under federal information-blocking rules.

3. The Right to Informed Consent

No procedure, medication, or treatment may be administered without your informed consent, except in documented emergencies. Informed consent requires that you understand what is being proposed, why it is being proposed, what the alternatives are, and what risks are involved. Consent given under duress, incomplete information, or impaired capacity is not valid consent.

4. The Right to Participate in Care Decisions

You have the right to participate actively in all decisions about your care, including the right to refuse treatment, request second opinions, and designate a healthcare proxy or advocate to act on your behalf. Your preferences, values, and goals of care must be respected and documented.

5. The Right to Advocate and Escalate

You and your family have the right to raise concerns about your care without fear of retaliation. This includes the right to request a change in your care team, to escalate concerns to the charge nurse, attending physician, or hospital administration, and to activate a Rapid Response Team (Condition H or equivalent) if you believe your condition is deteriorating and your concerns are not being addressed.

6. The Right to Transparency After Harm

If you are harmed by a medical error, you have the right to a full, honest, and timely disclosure of what happened, why it happened, and what the institution is doing to prevent it from happening again. This includes the right to participate in the review process and to receive fair compensation without being forced into adversarial litigation.

7. The Right to Privacy and Dignity

You have the right to have your personal information protected, your physical privacy respected, and your cultural, spiritual, and personal values honored throughout your care. You have the right to be treated as a person, not a case number, a diagnosis, or a revenue source.

8. The Right to Continuity of Care

You have the right to safe, standardized transitions of care during shift changes, department transfers, and discharge. This includes the right to have pending test results reviewed and communicated before you leave the hospital, and the right to receive clear, written discharge instructions that include warning signs requiring immediate medical attention.

APPENDIX B

Hospital Safety Checklist for Patients and Families

This checklist is designed for patients and their family members to use during any hospital stay. It is organized by care phase and is intended to be practical, actionable, and empowering. Keep it at the bedside and refer to it throughout the hospitalization.

At Admission

Identity Verification

Confirm that your wristband displays the correct name, date of birth, and medical record number. Verify this information with every new caregiver who enters your room. If anyone attempts to administer medication or perform a procedure without first checking your wristband, ask them to verify your identity.

Allergy and Medication Review

Provide a complete list of all medications you are currently taking, including over-the-counter drugs, supplements, and herbal remedies. Confirm that all known allergies are documented in your chart and on your wristband. Ask the admitting nurse to read your allergy list back to you.

Baseline Communication

Tell the nursing staff what 'normal' looks like for you or your loved one: typical breathing patterns, alertness level, skin color, pain tolerance, and any chronic conditions that might affect monitoring thresholds. This baseline information is invaluable and should be documented in the chart.

During the Stay

Medication Verification

Before accepting any medication, ask: What is this medication? What is it for? What is the dose? Are there side effects I should watch for? If the medication looks different from what you have been receiving, ask why before taking it.

Hand Hygiene

Every person who touches you or your loved one should wash their hands or use hand sanitizer first. You have the right to ask anyone, including physicians, to wash their hands before providing care. This is not rude. It is one of the single most effective measures for preventing hospital-acquired infections.

Shift Change Awareness

Be present during shift changes (typically 7:00 AM and 7:00 PM) whenever possible. Introduce yourself to the incoming nurse. Confirm that pending concerns, test results, and care instructions have been communicated. Ask: 'Is there anything from the previous shift I should know about?'

Monitor Awareness

Ask the nursing staff to explain what the bedside monitors are measuring and what the normal ranges are for your loved one. If an alarm sounds and no one responds within a few minutes, press the call button. If the call button is not answered promptly, go to the nursing station.

Daily Goals

Ask the care team each morning: What is the plan for today? What tests or procedures are scheduled? What results are we waiting for? What would need to happen for discharge to be considered? Write down the answers.

Before Discharge

Pending Results

Ask explicitly: Are there any test results still pending? If yes, who will review them and how will I be notified? Do not leave the hospital with unreviewed critical lab values. This single question could save your life.

Medication Reconciliation

Review your discharge medication list with the pharmacist or nurse. Confirm which medications are new, which have been changed, and which previous medications should be continued or stopped. Understand the purpose, dose, and timing of every medication on the list.

Warning Signs

Ask: What symptoms should prompt me to return to the emergency department? Get specific answers, not vague reassurances. Write them down. Know the difference between expected post-discharge discomfort and signs of a complication that requires immediate medical attention.

Follow-Up Plan

Confirm all follow-up appointments before leaving. Know who to call if you have questions after discharge. Request a direct phone number for the care team or a nurse advice line, not just the hospital's general switchboard.

APPENDIX C

Glossary of Key Terms

Adverse Event: An injury or harm to a patient that results from medical care rather than from the patient's underlying condition. Adverse events may be preventable (caused by error) or non-preventable (a known risk of appropriate treatment).

Alarm Fatigue: A condition in which clinical staff become desensitized to monitor alarms due to the overwhelming volume of false or non-actionable alerts, leading to delayed or absent responses to genuine emergencies.

Anchoring Bias: A cognitive bias in which a clinician fixates on an initial piece of information (such as a preliminary diagnosis) and interprets all subsequent data through that lens, potentially missing alternative explanations.

Automation Complacency: The tendency of human operators to become less vigilant when they believe an automated system is monitoring a situation, leading to delayed detection of anomalies or failures.

Availability Bias: A cognitive bias in which the likelihood of a diagnosis is judged by how easily similar cases come to mind, rather than by objective probability.

Black Box (Surgical): A recording system that captures video, audio, and physiological data during surgical procedures for non-punitive quality improvement and education, modeled on aviation flight data recorders.

CANDOR Process: Communication and Optimal Resolution; a framework for transparent disclosure, investigation, and resolution following medical errors, replacing adversarial litigation with collaborative accountability.

Closed-Loop Communication: A communication protocol requiring confirmation at every stage of information transfer: the sender transmits, the receiver reads back, and the sender confirms the readback is correct.

Condition H: A hospital policy that allows patients or family members to activate a Rapid Response Team directly when they believe the clinical staff is not adequately addressing a deteriorating condition.

Confirmation Bias: A cognitive bias in which a clinician selectively seeks information that supports an existing diagnosis while minimizing or dismissing contradictory evidence.

Diagnostic Error: A failure to establish an accurate and timely explanation of a patient's health problem, including missed, delayed, or wrong diagnoses.

EHR (Electronic Health Record): A digital system for storing, managing, and transmitting patient medical information, including clinical notes, lab results, medication orders, and imaging reports.

Handoff: The transfer of clinical responsibility for a patient from one provider or team to another, typically occurring during shift changes, department transfers, or discharge.

Near-Miss: An event that could have resulted in patient harm but did not, either by chance or through timely intervention. Near-misses are critical safety data because they reveal system vulnerabilities without resulting in injury.

PCA (Patient-Controlled Analgesia): A drug delivery system that allows patients to self-administer pre-programmed doses of pain medication (typically opioids) by pressing a button, with built-in lockout intervals to prevent overdose.

Premature Closure: A cognitive bias in which a clinician stops considering alternative diagnoses after arriving at a satisfactory initial explanation, without testing it against competing hypotheses.

Psychological Safety: A shared belief among team members that it is safe to take interpersonal risks—to ask questions, admit mistakes, and raise concerns—without fear of humiliation or punishment.

Rapid Response Team (RRT): A specialized team of critical care clinicians that can be summoned to any patient's bedside to assess and intervene when signs of clinical deterioration are identified.

Root Cause Analysis (RCA): A structured investigation method used to identify the underlying systemic factors that contributed to an adverse event, moving beyond individual blame to examine process, communication, and design failures.

SBAR: Situation, Background, Assessment, ecommendation; a standardized communication framework for clinical handoffs, adapted from the United States Navy submarine force.

Sentinel Event: An unexpected event involving death or serious physical or psychological injury, designated by the Joint Commission as requiring immediate investigation and response.

Sepsis: A life-threatening organ dysfunction caused by the body's dysregulated response to infection, requiring rapid identification and treatment for survival.

Silo Effect: The tendency of hospital departments to operate as self-contained units with separate information systems and communication protocols, creating dangerous gaps at interdepartmental boundaries.

Stop the Line: A safety concept adapted from Toyota manufacturing in which any team member, regardless of rank, has the authority and obligation to halt a process when a safety concern is identified.

Swiss Cheese Model: A framework developed by James Reason proposing that accidents result from the alignment of vulnerabilities ("holes") across multiple layers of defense, rather than from a single point of failure.

Troponin: A protein released into the bloodstream when the heart muscle is damaged; elevated Troponin levels are the gold-standard biomarker for diagnosing myocardial infarction (heart attack).